4374

500
FLA

Flatow, Ira.

Rainbows, curve
balls and other
wonders of the
natural world
explained

$15.45

RAINBOWS, CURVE BALLS

AND OTHER WONDERS OF THE NATURAL WORLD EXPLAINED

Also by Ira Flatow:

NEWTON'S APPLE
(General Communications Company
of America, 1983)

RAINBOWS, CURVE BALLS

AND OTHER WONDERS OF THE NATURAL WORLD EXPLAINED

Ira Flatow

4374

Illustrations by **Howard Coale**

William Morrow and Company, Inc.   New York

Library of Congress Cataloging-in-Publication Data

Flatow, Ira.
    Rainbows, Curve balls and other wonders of the natural
world explained / Ira Flatow; illustrations by Howard Coale.
        p.   cm.
    Includes index.
    ISBN 0-688-06705-0
    1. Science—Miscellanea.   I. Coale, Howard.   II. Title.
Q173.F6 1988   88-2654
500—dc19   CIP

Printed in the United States of America

First Edition

1 2 3 4 5 6 7 8 9 10

BOOK DESIGN BY BRIAN MOLLOY

To my father, Samuel,
who understood the joy and value
of learning, and to my dedicated
teachers, whose enthusiasm rubbed off
on their student

Foreword

Everything that is true is very simple,
once we understand it. It's only
complicated when we don't.
—Bernd Matthias, *American Physicist*

The most incomprehensible thing
about the universe is that it is
comprehensible.
                              —Albert Einstein

When my wife, Miriam, was pregnant, we faced a decision common to many expectant parents: Did we want to know the sex of our child in advance? Miriam thought about it for a while and decided she would rather not. I needed no time to think; I wanted to know immediately. Unfortunately for me, our amniocentesis experts said either both or none of us could be told. As you can guess, we were kept in the dark until Sam arrived.

When I tell this story to friends, they want to know why I didn't want to wait. Wouldn't knowing the baby's sex in advance take away the fun and excitement of that special moment? I replied that I enjoyed surprises as much as anyone but that the real fun in life is uncovering its secrets. Being kept in the dark about the world's mysteries is not my style. I want to know what makes things tick and why.

Einstein said it best: I want to know what God was thinking. Well, I'm no Einstein. I have no field of expertise in any branch of science, and while I have a degree in engineering, I have never practiced that profession. I've worked most of my life as a journalist covering science, health, medicine, and technology for radio and television. My view of the world is that of an inquisitive observer, lifelong learner, and skeptical inquirer. I share Einstein's curiosity and so do lots of others who seek and understand the rewards of knowledge.

Over the years thousands of people have written me seeking the answers to the mysteries that abound in science, nature, and technology. In this book I hope to share what I've learned, to answer some of the most popular and interesting questions that have been asked me over and over again. This is not a textbook. It does not explore each topic in full. I leave that to others. And unlike other science books that organize material by topics, I've devoted each chapter to a real situation—going to the beach, cooking in the kitchen, spending a night at the theater, etc.—that lends itself to investigation. I think learning is more enjoyable when we actively experience a learning situation. In the future, you might wish to consider a concert hall and your shower stall an ongoing experiment in sound or your kitchen a culinary laboratory.

It may sound odd, but if nothing else, I hope this book brings a smile to your face and makes you nod your head in recognition. Because then I know you're saying to yourself, "Isn't that interesting. I never understood that before."

Stamford, Connecticut   —IRA FLATOW
February 1988

Acknowledgments

I am grateful to many hardworking people who knowingly or unwittingly helped make this book possible. My thanks to the producers, researchers and expert guests at *Newton's Apple,* who, over the years, did much of the digging that uncovered many of the facts found in this book. Special thanks to Leslie Kratz and Marie Domingo, who cheerfully supplied the answers to last-minute requests.

I've been blessed knowing people who are not only experts in their chosen fields but in love with what they do and need no prodding to share their knowledge. Much of their enthusiasm and drive is contained between these covers. Little of the astronomy in this book would have been possible without Dr. Lawrence Rudnick, friend and physicist at the University of Minnesota, who spent many hours with me discussing the hidden joy and meaning of physics. My thanks to Dr. Jan Serie, biologist at Macalester College in St. Paul, Minnesota, who demystified the workings of the human body with the help of a skeleton; to Dr. Jeffrey Hoffman, NASA astronaut, physicist, friend, and perhaps the first Renaissance man to walk in space; to Dr. Vance Vicente, marine biologist at the University of Puerto Rico in Mayaguez, who opened my eyes to the secrets of sand.

When I wanted to talk food, I turned to Dr. Art Grosser, indefatigable chemist, actor, and author of *The Cookbook Decoder.* Dr. Howard Brody of the University of Pennsylvania was just as eager to talk sports as was George Manning at Hillerick & Bradsby Company.

Dr. Daniel Keyler, of the Hennepin County Medical Center in Minneapolis, was more than eager to discuss snake bites, and Dr. George Bissinger, East Carolina University, Greenville, North Carolina, gave me my first lesson in singing in the shower. Many thanks to Dr. E. Philip Krider, director of the Institute of Atmospheric Sciences, University of Arizona, and Fred Gadomsky, meteorologist, Penn State University, for their help unraveling the weather. My

9

appreciation to Dr. Maria Hordinsky, Dr. Jearl Walker, and Dr. Cyril Harris for their thoughtful guidance.

My thanks, too, to Maria Guarnaschelli, senior editor at William Morrow and Company, for her patience in dealing with a nervous author. I am also grateful to Howard Coale, whose unique illustrations adorn these pages. This book would not have been attempted without the prodding of Bill Adler.

My wife, Miriam, has been more than generous, sacrificing her own time and taking over many of my duties so I could write.

And finally, though it may sound odd, I could never have finished the book without the undying support offered by new technology: my pc-clone computer and word processors NewWord and WordPerfect. For someone who can't write three words without making four typos, the delete key is a godsend.

Contents

Contents

I. By the Beautiful Sea

The shore is an ancient world,
for as long as there has been an earth
and sea there has been this place
of the meeting of land and water.
—Rachel Carson, *The Edge of the Sea*

Skeletons in the Sand

Illustration 1

Here you are lolling on a tropical beach, sunglasses fixed snugly, body covered with lotion, a good junk novel waiting to be read. As thoughts about the office and the boss melt away, the mind turns to nature. As far as the eye can see, soft, white beach sand covers your tropical paradise, and for thousands, millions of square miles around, sand or clay blankets the bottom of the sea. But where did it all come from? How did it get here?

Like many people, I had the impression that beach sand started out as a big rock caught in the pulverizing clutches of the waves; that sand was produced by the ceaseless pounding of ocean on stone, cracking and smashing the boulders into small pebbles, grinding them into grains of sand. How else could one account for the universal presence of sand at every seashore?

But I was wrong. True, the kind of sand found when digging a hole to build a house or bury a bone was formed that way: The erosion of granite or granitelike rock produced the typical gray sand found in the backyard. But for the beautiful white sandy beaches on many tropical shores, sand had a much different beginning: living beings.

Many if not most of the grains of sand on these beaches were part of living things. These grains are the bony, glassy remains of marine plants and animals that were once alive and in their death leave us sand. In other words, white sandy beaches are graveyards of ancient marine organisms.

The sand may be of calcareous origin, which means that it is the

15

hard remnant of an organism that used calcium as part of its everyday routine: as a structural support for itself or as a shelter that served as protection. Or the sand may be the skeleton of an organism that lived in a house of glass.

For example, living in the coral reefs of the Caribbean are plants called coralline algae. Drawing calcium from seawater, these algae secrete a hard, chalky substance—calcite—that helps to build coral reefs. When the algae die, they leave behind their hard calcium carbonate bodies, which may be ground up by other organisms, producing chalky chips that help make up sand.

Ocean-dwelling single-celled animals called foraminifera leave behind hard, calcite shells.

Glass Houses

Marine animals whose bodies secrete glass may be large or small. Single-celled animals called radiolaria have skeletons of glass (silica), which is a common element of the clay of the deep oceanic floor. Look inside the world's largest sponge—the loggerhead—and you'll see a creature full of glass. Tiny pointed "spicules" of glass are made by the sponge as it absorbs silicates from the seawater. With the help of cells called sclerocytes (sclero = hard; cytes = cells), the sponge deposits the silicates as glass crystals, which look a lot like the layers of fiberglass that run through an attic or like glass belts on a tire. In fact, about half the bulk of a three-foot loggerhead sponge is made of glass. When mixed with fragments of dead coral, snails, sea-urchin spines, foraminifera, and other types of sponges, these remains form that nice, white sandy beach you love to burn your feet on.

Plants also leave behind glassy remains. Diatoms, one-celled plants with hard silica skeletons, live in the surface waters of all oceans but are found mostly in the cold waters of the North Pacific and Antarctic. They extract silica from ocean water and create glasslike shells. They are so proficient at this that when diatoms grown in a laboratory aquarium use up the readily available silica in the water, they will begin to take silica from the glass walls of the aquarium. How they accomplish this is not clear. When diatoms die, their glassy shells, or frustules, sink to the bottom of the sea, creating a wide swath of debris on the ocean floor. When the debris hardens into rock,

Illustration 2
The loggerhead sponge is the largest sponge known. It's been
reported that one loggerhead, when cut open, was found to
have sixteen thousand animals, mostly shrimp, living in its
canal system.

it is called diatomite. One of the most famous and accessible
diatomites is the Monterey Formation. It can be viewed along the
coast of central and Southern California.

The production of sand by living things is made even more
remarkable when one considers the global scale on which this occurs.
It's hard to imagine the number of years and countless living creatures

Illustration 3
Diatoms

17

that died to produce the white sandy beaches of the world. Is there a number that large? And these beaches are but a fraction of the unseen remnants of past life. Far greater evidence lies hidden, covered by ocean and mountaintop.

The limestone we use today comes from layer upon layer of plant skeletons built up over hundreds of millions of years. The Big Belt Mountains of Montana—containing six thousand square miles of limestone—were laid down about five hundred million years ago by ancient algae. (The multicolored deposits called tufas at the base of the hot springs in Yellowstone National Park are believed to have been made by relatives of these algae.) On an even larger scale the "ooze" that coats the bottoms of almost all the seas are the remains of one-celled plants and animals.

The shell remains of single-celled animals who died three hundred million years ago can be found in the limestone and rock used to build the Egyptian pyramids. The famous White Cliffs of Dover are composed of calcium chalk. But look under a microscope and you'll see that the sand is made from the one-celled marine animal called foraminifera (foram). Foram skeletons can be found almost everywhere. They cover the bottom in all the oceans except the coldest, Arctic and Antarctic. These deposits have been laid down over millions of years, and in some places in the Atlantic basin sediments up to twelve thousand feet thick have been found.

Black, White, Pink, and Green Beaches

A colorful species of foram, rubrum (rubrum-red), is responsible for the rosy sand found on pink beaches in the tropics. Of course, *all* sand at the beach was not made by living things. Next time you stroll along the shore, look for wavy lines of black among the white. Black sand is the product of eroded rock first blown out of a volcano centuries ago. Some Hawaiian beaches, such as Kalahana Beach, are the result of lava flowing into the ocean where it hardens into solid black rock. Crashing waves break up the rocks into tiny particles. Over the years these sandlike particles are washed up along the shores of white beaches near the volcano. Eventually the black sand covers and replaces the white calcite sand. Place a magnet on the black sand and black particles will be pulled away. Iron is an important component of black sand.

Beaches with black sand are rare, but Hawaii also offers a few "green" beaches. Grains of an emerald green mineral called olivine make up these beaches. Some lava flows may contain as much as 40 percent olivine, but such high accumulations are very rare. Green beaches are scarce, and the few that exist can be found at South Point, Hawaii.

Illustration 4

The endless miles of desert sand dunes you see in *Lawrence of Arabia* exist only in the minds of Hollywood writers. Drifting sand covers only about one seventh of the Sahara Desert, and about a third of the Arabian Desert. On the average, only about one tenth of all desert areas around the world have these legendary sand dunes. Sand in these dunes is made mostly of the mineral quartz, the same stuff that keeps accurate time in your digital watch. Quartz is among the most abundant minerals on earth (and this abundance is partly why quartz watches are so cheap).

· ·

TRIVIA: The city of Richmond, Virginia, rests on a bed of "diatomaceous earth" fifty feet thick. Diatomaceous earth is really the skeletons of tiny one-celled plants called diatoms. Because the shells of diatoms are literally made of sharp glass, diatomaceous earth is widely used commercially as an abrasive. You may be brushing your teeth with it. Winemakers use it to filter out yeast cells and remnants of fermented grapes. One cubic inch of diatomaceous earth may contain as many as forty million shells.

· ·

The Big Burp

You can live with someone your whole life and never really get to know him or her. The same thing can be said about our constant relationship to the substance water. Water is everywhere: we splash, wash, and walk in it; drink, drown in, and skate on it. When there's not enough, we pray for rain. When there's too much, we prepare for a flood. Water can be hard one minute and soft the next. But to truly know and appreciate the versatility of water, you have to look deeply inside its structure, and understand what makes a water molecule tick. There you'll find a unique shape and composition that give water seemingly magical powers, enough to impress even Houdini (who, incidentally, used water often in his act).

A Refreshing Antique

Try pouring yourself a fresh glass of water. You can't. Age is meaningless in relationship to water: Just about every drop of water is ancient. The water in that glass of iced tea or in that cup of coffee may taste fresh, but it's been around for *billions* of years. Billions. *Fresh* is a ridiculous concept when applied to water. It's believed that all the water on the earth today was created in one short geological period when the earth was young, when hot gas from inside the earth erupted through volcanoes, geysers, and hot springs—what scientists call "the Big Burp." When the earth cooled to just below 212 degrees Fahrenheit, this water condensed into all the oceans, lakes, and ice caps we see around us on the earth today. So just about all the water you see on the earth was released by volcanic action about 4.5 billion years ago, just a few hundred million years after the formation of the earth, and has been recycled ever since through rain and evaporation. Antique collectors can find no better prize than a bottle of seltzer.

Could Comets Have Created the Oceans?

Another theory proposes that the bulk of the water on the earth comes not from the earth's center but from outer space, from countless numbers of comets that crashed into the earth billions of years ago. Comets—playfully called "dirty snowballs" by scientists—are mostly ice. Though not widely accepted, the comet theory is based on a view that comets showered our planet early in its lifetime. A rain of comets lasting for billions of years might have supplied enough moisture to create the oceans we see today.

Almost all of the world's ice (90 percent) and most of its fresh water is locked up at the bottom of the world, in Antarctica. The Antarctic ice cap contains about 7.2 million cubic miles of ice, holding 70 percent of the world's store of fresh water. Each year about 1.4 trillion tons of ice break off and melt into the southern oceans. Tabular icebergs measuring a hundred square miles commonly break off from the ice shelves. A single iceberg the size of Luxembourg once broke away and was set adrift in the Weddell Sea. These flat, or tabular, icebergs, shaped

Illustration 5

Could comets have caused our oceans? One theory states that over a period of many years, a shower of comets—dirty "snowballs" from outer space—may be responsible for bringing the water to earth that now fills our oceans.

like stepping-stones, are much different from the saw-toothed icebergs spawned from glaciers in the northern, Arctic regions. If the earth's climate were to change and both polar ice caps melt, the water released would raise the level of the world's oceans by about two hundred feet.

Amazingly enough, over all these billions of years, the amount of water on the earth has remained nearly constant. Water molecules are too heavy to escape gravity and float off en masse into space, but every so often a few will dissociate (break up into hydrogen and oxygen). Then the very light hydrogen can and does escape. But the amount of water escaping into space each year is very little—equal to about two hundred million gallons. Over the life of the earth all the water lost into space would fill an ocean about eight hundred square miles. But there is no need to worry. The oceans are not slowly vanishing because as the water vapor escapes, new water called "juvenile water" is replacing it. Juvenile water seeps up from the interior of the earth via cracks in the earth's crust. It is not *new* water in the sense that it hasn't just been created. Rather, juvenile water has not previously existed above the earth's surface. It may find its way up through cracks in the ocean floor. Or be part of the water that spurts out of a geyser or hot spring.

The Head of a Teddy Bear

The genesis and history of water is only the beginning of an amazing story. To a chemist or a physicist, a biologist or an engineer, water is truly the most marvelous and versatile substance on earth. Why? Mostly because of the way it looks. Water has unique and surprising powers due to its lopsided chemical structures which we all know as H_2O. If you could see a water molecule under a microscope, it would look a lot like the head of a teddy bear or Mickey Mouse. Two atoms of hydrogen form the ears, and the oxygen atom the face. This simple molecular structure (kind of a cuddly appearance) is deceiving; it gives water unique powers. The hydrogen atoms sticking out of one side make water molecules behave like little magnets: The "ears" are positively charged, the face negatively charged. The magnetic character of water gives it an incredible ability to dissolve substances. Water literally yanks matter out of a solid and keeps it in solution. Picture what happens to a lump of sugar in a cup of coffee: The forceful pull of the powerful water molecules rip the weaker bonds that hold

23

the sugar together. Water molecules wedge them apart until every sugar molecule is surrounded by water—dissolved.

Water's magnetic behavior is the reason H_2O is called the universal solvent. Given enough time, water will dissolve any organic substance on or above earth—dissolved chemicals in the sky bring acid rain, dissolved minerals on the ground fertilize plants and feed

Illustration 6

animals. Water is so busy dissolving substances that you'd be hard pressed to find "clean" water, free of dissolved particles, anywhere outside a laboratory. The steam coming out of your tea kettle comes close.

Equally surprising is how strangely water molecules react to one another. Each molecule of H_2O can link with four others. When water is frozen, these links form rigid, crystalline patterns. But in its liquid state, water is in constant motion. The molecular links are made and broken, then reconnected and broken again—a flurry of chaotic activity in a substance that appears to be still. It's the "average" of all this action that we perceive as a "peaceful" glass of water. Yet amid all this turmoil, the bonds remain strong enough to support a razor blade laid gently on the surface—a property called surface tension.

Water's love affair with itself—its attraction for its own molecules—serves nature's needs and helps make life possible. For example, consider this problem: How does a four-hundred-foot Douglas fir tree get water to its uppermost branches? It has no electric or mechanical pump. But it doesn't need one because water has something better going for it: Water can defy gravity.

Hydrogen atoms in water will cling not only to oxygen atoms of other water molecules but will also adhere to oxygen atoms of any other material—animal, vegetable, or mineral—in a process called capillary action. A water molecule that starts in the soil will cling to oxygen-laden roots, where it meets another water molecule. One will grab hold of the other and become part of a long chain of water molecules stretching hundreds of feet to the top of the plant, as in

24

Illustration 7

Illustration 7. As a water molecule evaporates off a leaf at the top—transpiration—another one is pulled up the chain to take its place. The immense strength of the molecular bonds—cohesion—holds the water molecules locked together, and the chain remains unbroken. By this capillary action the plant moves water and nutrients through its system of long, thin tubes we call wood. And by the same capillary action water makes life possible for animals as well.

Our bodies contain sixty thousand miles of arteries and veins, soaking our tissues with oxygen-carrying blood and nutrients. The tiniest of the blood vessels, the capillaries, get their name from the action of blood moving through these microscopic channels. In other words, our bodies have evolved a plumbing system uniquely designed to take advantage of one of water's wondrous workings. To keep our systems going, each of us must consume almost seven thousand gallons of water in a lifetime. Lose just 2 percent and we get very thirsty. A loss of 7 percent and circulation shuts down, resulting in death.

Missing Ice Cubes

Water behaves even more strangely when cold. How many times have you filled the ice trays only to find the ice has frozen over the top or shrunken almost to nothing? It all depends on how long you've let the ice lie in the freezer.

Most substances, when they cool off, contract. So does water—up to a point. That point is about 39 degrees Fahrenheit. From there, as it gets colder, water does something unusual for liquids: It expands. That's why your ice cubes are much larger than the cubicles they froze in. Odd as it may seem to us, this property is extremely useful to nature. It explains why lakes and rivers freeze only on the surface. At 39 degrees, right before it begins to freeze (expand), water is most dense, so it sinks to the bottom of the lake. Ice, being lighter, rises to the surface. If this were different, lakes would freeze from the bottom up and remain frozen throughout the summer. Life could not exist on Earth.

In the process of expanding, water exerts a tremendous force. That's why automobile engine blocks crack when they freeze (don't forget the antifreeze) or forgotten soda bottles shatter in the freezer. When water seeps into cracks in the sidewalk or highways, it can wreak havoc in the winter by splitting open the pavement. It's not the salt on the pavement that cracks concrete. It's the liquid water, melted by the salt, that refreezes later.

As for those tiny ice cubes that used to be quite large, they were left in the freezer too long: The water vaporized. But wait a minute. Doesn't water have to be liquid before it can change to a gas? No. Water can change from a solid to a gas without having to melt first. This is called *sublimation*. That's what happens in the freezer when large ice cubes are reduced to lowly ice chips over a period of weeks. There's a wonderful place in Antarctica called the Dry Valleys, given that name because though the ground is always frozen, it is ice-free. If it weren't for the cold, you might think you were in a peaceful California desert. Why no ice? Because winds of up to two hundred miles per hour come by regularly and blow away the snow. The bluster speeds up sublimation, leaving the ground ice free.

Ice itself is quite fascinating. Ice is held together by strong chemical bonds that give it a crystalline form. So ice is not only hard as a rock, in a sense, it really is a "rock"—a mineral. That makes a glacier a large rock made of countless numbers of ice crystals looking a lot like snow. Snow is ice crystals displaying an infinite variety of shapes.

If ice is really a mineral, how can you skate on it? You can't skate on a rock. Why can you skate on ice? A fascinating (and very useful) property of ice is that when you press on it, ice melts. So when the weight of your body presses on the ice, the melting ice forms a tiny film of water under the skate, which lubricates the skate and makes the ice very slippery. People who measure the slipperiness of things say that water-soaked ice is the most slippery substance in the world.

Ice is not unique to our planet. Water ice caps cover the poles of Mars. Although no ice has been discovered on the moon, the "moons" of many of the other planets in our solar system are loaded with frozen water. The rings of Saturn contain chunks of water ice of varying sizes, dubbed "snowballs" by astronomers. None of these places is hospitable enough to support life that would need water to survive.

• •

TRIVIA: Even during the coldest months of winter, the water at the bottom of deep lakes never freezes. Instead, ice floats to the top and only the surface freezes. If lakes froze to the bottom, fish and other life forms would be killed. The water below remains at 39 degrees Fahrenheit. That's why knowledgeable campers pitch their tents on frozen lakes: They know the water below the ice is warmer than the frozen ground and that the warmth will radiate upward.

• •

Crests and Troughs

As a little boy, I used to stand on the seashore and wonder where in England the waves were coming from. Was there another little boy standing someplace across the Atlantic wondering the same thing about America?

Of course, this question assumed that the waves originated on opposite sides of the ocean, as if the landmasses of Europe and North America were quivering, setting up little ripples that eventually pound on each other's shores.

The mystery of where waves originate is compounded by the observation that they just keep on coming. They *never* stop. Day and night, fair weather and foul, the waves just keep on rollin' like they've been at it forever. It's the closest thing to perpetual motion we know. But why?

The Answer Is Blowin' in the Wind

Waves can be generated in a lake or swimming pool in one of two ways. You can drop a pebble into the water and watch the concentric wave pattern that spreads from the point of impact. Or you can wait for moving air—wind—to initiate waves. Both methods are at work in causing wave formation in the oceans.

The most common waves are caused by the wind. You could no more stop the wind from blowing than the waves from rolling. Wave size depends on wind speed, wind duration, and fetch, the distance of water over which the wind blows. The longer the distance the wind travels, the higher the waves will be. The greater the wind speed, the more powerful the waves.

As the wind begins to blow, it tries to drag the ocean surface with it. The surface can't move as fast as the air, so instead it rises. But, of course, gravity steps in, grabs hold, and brings the water back down. The momentum of the falling water carries it below the surface.

Water pressure from below pushes it back up again. The resulting up-and-down seesawing, this tug-of-war between gravity and water pressure, is the essence of wave motion.

With a breeze of less than two knots, the air causes nothing more than tiny little ripples called capillary waves. If you've ever seen a sudden gust of wind send ripples across a quiet lake, then you've seen capillary waves. When the ripples reach about two-thirds of an inch long, they are strong enough to overcome the surface tension of the water and become true waves, albeit tiny ones. As the wind gathers strength, it blows hard enough to cause these waves to combine at random, forming longer and shorter waves and creating a choppy sea.

Crank up the wind speed a little more, to about thirteen knots (fifteen miles per hour) and now the waves grow taller faster than they grow longer until their steepness causes them to break. These are called **whitecaps.** For a whitecap to form, the wave height must be one seventh of the distance between wave crests. This distance is known as the wavelength. Therefore a wave measuring seven feet in wavelength (crest to crest) will break when it's a foot high.

So far in this scenario, no long ocean swells are rolling toward the shore. Just a jumble of choppy whitecaps. Whitecaps have no sustained energy; they just collapse into a hodgepodge of turbulent water. Yet out of this confusion comes order. Longer waves, by nature, move faster. The longest waves leave the windy area quickest, traveling together and combining to form swells that do reach the shore. The first swells to reach our shores belong to the long wave groups. Some of the shorter waves die out en route.

A wave may travel thousands of miles before it reaches shore. Out at sea when the wind stops blowing, a wave may be reduced to nothing more than a nice, gentle swell. But once it approaches land, all that changes. The shoreline is shallow, so the base of an approaching wave drags against the bottom and is slowed down. The wave crest, which continues to move quickly, curls over and breaks. A wave breaks when the water at the crest moves faster than the water supporting it below. This precise condition occurs when the depth of the water is about 1⅓ times the height of the wave. A four-foot wave will break when it reaches shallow water of about

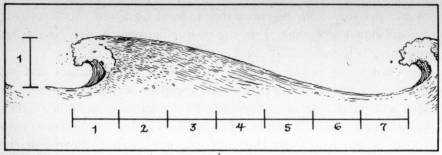

Illustration 8

5¼ feet. A forty-foot wave may break far offshore, forming smaller waves that break themselves, etc. Sandbars or an irregular ocean bottom a few hundred yards off the beach may cause the waves to break and reform several times on their way to becoming surf. What makes the surfing great at places like Waikiki Beach is the almost flat bottom and long Pacific waves—ideal conditions for long stretches of hollow, tubular breakers.

As waves approach the coastline, the wave front will change to match the shape of the underwater geography. The crests of waves in shallow water move slower than those in deep water, causing the wave front to bend. If the bottom is smooth and equally deep in all directions on the shoreline, the waves will bend to become parallel with the shore. You can get some idea of the underwater features of the coast by carefully studying the incoming waves at the beach.

What about those monster tidal waves, the kind you see knocking over ocean liners in the movies? Leave them to Hollywood. Tidal waves, more correctly called **seismic** waves or *tsunamis,* would hardly be noticed at sea. They seldom exceed one or two feet in height. Tidal waves are actually caused not by wind but by

Illustration 9

30

Illustration 10

earthquakes or volcanoes. They result from the violent shaking of the water by earthquakes or volcanic eruptions. (The creation of this wave is analogous to our example of dropping a pebble in the swimming pool. While not as tumultuous as an earthquake or volcano, a pebble can create enough of a disturbance to set up wave motion.) Earthquakes and volcanoes produce such long waves, a hundred to two hundred miles long, that the energy of the wave is carried in its length, not its height at sea. Although a tsunami may travel at five hundred miles per hour, its real destructive power comes when it approaches shallow water and its wavelength is reduced. Then a small swell can grow to a hundred-foot-high wave and wipe out an entire coastline. A tidal wave following the volcanic eruption at Krakatoa, Indonesia, in 1883 killed thirty-six thousand people in nearby coastal waters.

It doesn't pay to take chances with waves, even smaller ones. If you've ever been hit by a wave or have watched a hurricane rip the sand from a beach, you know how strong waves can be. Waves have reportedly moved twenty-six-hundred-ton blocks of stone and thrown hundred-pound rocks into the air.

"Rogue" waves produce those huge oceanic titans that tower over ships and crash them to bits. Rogue waves are freak occurrences produced when many waves converge by chance on one spot. A group of rogue waves pounded a fleet of racing yachts participating in the Fastnet Race off the coast of Great Britain in August 1979. A series of monster waves fifty feet high, demolished twenty-three boats, killing fifteen people. The waves were made by a single storm in the area, producing waves from two different directions as it changed course.

31

The waves arrived at the same time from remote spots and converged to produce the powerful rogue waves.

Despite their power, one of the fascinating things about waves is that they don't actually move much water. Watch an ocean wave carefully and the first thing you'll notice is that the bubbles floating on the water don't go anywhere. They just move up and down in the swell. You can also observe this by riding the waves out past the break line. The swells will carry you up and down but the water itself won't move forward until the wave breaks. The constant march of waves to the shore is really an optical illusion. Only at the very end of a wave's life, just as it's forming into a breaking crest on the beach, does the water move forward. If anything, say scientists, the water in a wave at sea tends to move in a circle. Water moves upward on the front edge of the wave and downward on the back. The water is not displaced from where it was, but like a roller coaster ride, moves in a circle.

· ·

TRIVIA: The tallest wave ever observed was 112 feet high. On February 7, 1933, a naval officer on the bridge of the tanker U.S.S. *Ramapo* spotted the giant in the North Pacific. The seaman kept his cool long enough to record the event despite the towering ten-story wave looming before him. It is the largest deep-water wave ever authenticated.

The rough surface of the ocean is called "sea." The choppy texture comes from the interaction of many waves of different heights, direction, and wavelength when they meet.

· ·

When the Air Has Color

Into my heart an air that kills
　　From yon far country blows:
What are those blue remembered hills,
　　What spires, what farms are those?
—A. E. Housman, *A Shropshire Lad XL*

Leonardo da Vinci, master of art and science, was one of the first people to document how atmosphere affects color. As a painter with a keen eye toward subtle shades of coloration, Leonardo noticed that objects changed color with distance. The farther away a mountain was, the bluer its appearance. He knew that the thickness of the layer of air between the observer and the object caused the color shift to blue, but he didn't know why. He didn't understand that the sunlight was in effect making the mountains appear behind a filter of blue. Poet A. E. Housman also recorded the mysterious bluish mountain tint in his poetry. Both men never realized that blue mountains and blue sky are all due to the same principle: light scattering.

Despite popular conceptions, the sky is not blue everywhere. When you look at pictures of outer space you don't see the Space Shuttle flying through a sea of blue. The astronauts bouncing on the moon were not silhouetted against an azure sky. What you do see is a lot of black. It's only on earth that the sky appears blue.

That's because our planet has something that doesn't exist on the moon or in outer space: an atmosphere. It's the sunlight interacting with the atmosphere—the air—that makes the sky blue.

Remember that sunlight is composed of a rainbow of colors. Isaac Newton was the first to show that a prism does not create beautiful colors but merely separates them out of white sunlight. Somehow the air filters the blue light out of the white and makes it visible to us. It does this by the nature of the particles that make up the air: the air gasses, water, and dust.

Simply stated, air particles scatter sunlight. Scattering is the deflection of light rays by fine particles. Scattering is quite obvious in a smoke-filled room. When sunlight enters through a window it is scattered by the particles of smoke, resulting in visible shafts of light. Light comes in all different sizes called wavelengths. Each color corresponds to a different wavelength. Red light is long. Blue light is short. When there are particles in the atmosphere of a size much smaller than the wavelengths of the colors, selective scattering occurs: The particles will scatter one of the colors, and the atmosphere will appear to be that color. Shorter wavelengths (blue) are scattered much more strongly than long ones (red). So as the sunlight goes through the atmosphere some of it, especially the blue, bounces off the air particles and becomes visible to our eyes. Technically we say that the air scatters the blue light and produces the blue sky.

At sunset the sky changes color for the same reason. As the sun drops toward the horizon, the sunlight has more atmosphere to pass through, as seen in Illustration 11. The sunlight loses more of its blue

Illustration 11

rays as it slices a longer path through the air. What's left are the colors at the other end of the light spectrum—orange and red—the stuff of magnificent sunsets. The sunset can be made even more dramatic by the presence of clouds. The tiny water droplets in clouds are very good at scattering out the "cold" colors, leaving behind a pink sky. Volcanic eruptions have produced some of the most spectacular sunsets of all times. The fine dust particles sent aloft and carried around the world serve to scatter the sun's rays, producing beautiful and dramatic sunset colors. Ironically, air pollution, while contaminating the atmosphere with noxious gases and particulate matter, in the same manner has helped produce striking sunsets.

In effect, the air has real color. It is filled with scattered sunlight. Anyone who's wondered why mounds of snow or the ocean appear blue now know it's the air itself that gives objects their color. The veil of blue is still a delight to tourists who visit the Blue Ridge Mountains of Virginia and to gourmet coffee drinkers whose beans grow in the Blue Mountains of Jamaica. Oregon and Maine as well as Australia and a dozen other places around the world have parts of their geography named "Blue Mountains" in recognition of the artistic effects of nature.

Illustration 12

• •

TRIVIA: The sea is blue for the same reason as the sky: the scattering of light. The molecules of seawater scatter out the blue color component of white sunlight.

• •

Why the Sun Tans Your Skin but Bleaches Your Hair

The reason your skin gets dark and your hair gets light is because your hair is dead and your skin is alive.

The business end of the hair is at the follicle, under the top layer of skin. It's in the root where a hair begins to form. Supplied by blood vessels, the root is alive and as the hair grows, the root gives off a protein material into the shaft. By the time the hair gets to the surface of the skin, it is completely made of protein.

While this protein is being secreted to produce hair, there are pigment cells that are putting in the color. It's a dark brown pigment, melanin, that makes a brunette dark-haired.

As the hair grows, the melanin reaches the surface of the skin. When the melanin meets the bright sunlight, a chemical reaction takes place that converts the dark pigment to a clear color. This bleaching is time-dependent. So the longer your hair is exposed to sunlight, the clearer it gets.

While the sun is bleaching the melanin in your hair, it is tanning the melanin in your skin. This process occurs because your hair is dead but your skin is alive. This is a section of skin shown before exposure to the sun. It's made of many layers. A dead layer seals the top, while at the bottom more pigment cells are found.

What happens when you lie on the beach, exposed to the sun? Radiation from the sun begins to warm your body and make you feel good. But your skin panics. Sunlight may be great for your ego but it's lousy for your skin. Long exposure to the sun's ultraviolet (UV) light can degrade your skin and possibly cause cancer.

Even in a mild dose, ultraviolet begins to hurt the "prickle" cells in your skin. The blood vessels swell with blood. The skin's sensory receptors are damaged and say, "Ouch!" when touched.

36

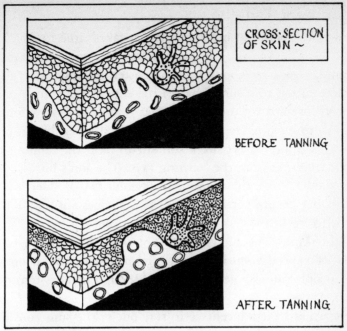

Illustration 13

The symptoms are familiar. At this point you have a mild sunburn. Soak up lots more sun and your skin starts sending up fluid and white blood cells to protect the area. Your sunburn has advanced to the blister stage—a second-degree burn.

To protect itself from burning and damage the skin rallies its forces. It begins by thickening its top dead layer (see Illustration 13). Meanwhile, pigment cells down at the lower skin level rev up to make more pigment, more melanin. The game plan is to make enough tough pigment to absorb the ultraviolet light and protect you.

Gradually the pigment gets spread throughout all the cells. As it spreads toward the top layers and into the prickle cells, the pigment gets darker. A tan is born. A tan is the skin's natural defense against the harmful rays of the sun. Melanin is an excellent sun block. Not only does it absorb UV radiation, but also some of the pigment forms a cap over the living cells in the dermis, protecting the genetic material in the skin. Skin cancer develops when the UV breaks the strands of the cells' DNA.

37

Timing a Tan

It's worth pointing out that the tan takes time to develop. The pigment takes days to work its way up to where you can see it. On the other hand, a sunburn shows up in a matter of hours. That's why it's a common notion to assume that in the course of a couple of days, a sunburn turns into a suntan. But that notion is dead wrong. That process—turning pink to brown—works well when broiling a steak but does not apply to getting a suntan. In fact, the tanning rays are distinctly different from the burning rays. The sun's burning ultraviolet is called UV-A, the tanning ultraviolet UV-B. Getting a good burn does not ensure getting a good tan. It does ensure a lot of pain and an increase in the risk of getting skin cancer.

Light-skinned people can get sunburned in only ten to twenty minutes of exposure during peak sun hours around noon. Black-skinned people can also get sunburned, though it takes ten to twenty times the exposure.

If you need extra incentive to stay out of the sun, consider this: Not only does extended exposure raise your chances of skin cancer, but sunlight also makes your skin wrinkle and age prematurely. Sun damage is cumulative. Over time the sun breaks down the elastic material in the skin, replacing it with a substance called elastone. Elastone is a poor substitute for the skin's resilient foundation; it does not hold the skin tightly. The result: sagging, drooping, and wrinkled skin.

What should one do? Stay out of the sun; or if being a shut-in is not your lifestyle, try a sun block. A sunscreen rated number 15 or higher should allow you to stay outside as long as you want without danger to your skin.

• •

TRIVIA: Don't be fooled by cloudy weather. Clouds may absorb the infrared or heat rays of the sun, but ultraviolet radiation still reaches the skin. Long exposure, even in the shade of an umbrella, can cause sunburn from scattered and reflected radiation.

Tanning reaches a peak in four to ten days. Fading occurs when pigmented skin gradually sheds. But once skin is sunburned, the damage cannot be undone. The effects are cumulative. Prolonged exposure to UV disrupts the structure of the skin, makes it lose elasticity, causes premature wrinkling, and increases your chances of getting skin cancer.

. . . .

Thunderbolts of Zeus

Illustration 14

If you linger on the beach into late afternoon and night, you may be lucky enough to catch a magnificent show: lightning.

I'm not suggesting that you stay put and let lightning melt your toes off (lightning kills a hundred people each year in the United States). Get to shelter if a storm is close by. But many times you may be able to see thundershowers out at sea and safely watch lightning dance between clouds in the distance. It is an amazing sight and one you will not forget. I remember watching such an event on a beach in Florida and being riveted by the light show the storm put on. Jagged streaks of lightning jumping from cloud to cloud lit up the night like cannon shot in the distance. The ancient mythological gods Zeus, Jupiter, and Thor regularly kept order by hurling lightning bolts from the sky.

Thunderstorms are so beautiful that a little insight into their mechanics might dispel the fears of the most timid among us and reveal a thunderstorm for what it is: a magnificent demonstration of electrical physics.

Science has come a long way since Ben Franklin's famous kite-flying experiment in 1752, but the mechanism that produces lightning is still not understood very well. Franklin discovered that the small sparks set off in the key attached to his kite string were identical to the sparks he could produce in his laboratory. So in effect

40

Franklin discovered that lightning was really electricity, a giant spark. We now know that this bolt is a cloud's way of releasing pent-up electrical charges produced during a storm. Since a thunderstorm is merely a swirling mass of wind, ice, and raindrops, it's not clearly

Illustration 15

understood how the electrical charges that cause the bolt are produced. What we do know is that for a lightning bolt to be created, there has to be a tremendous buildup of charge. Positive and negative charges must be separated so they can be discharged suddenly. Collisions between ice crystals and hail particles in the presence of supercooled water drops (water still liquid below freezing temperatures) may contribute to the separation of the charges. Wind currents within the cloud distribute the charges. They collect charges building up on cloud boundaries, bring them inside, and concentrate and make the electric field larger. In a thunderstorm the upper part of the thunderclouds is full of positive charges while the lower part is packed with negative ones; see Illustration 15.

Once the charge difference becomes great enough, the insulating air's natural resistance breaks down. It cannot keep the charges within clouds and between cloud and ground separated. A flash of lightning jumps within a cloud or between cloud and ground. (Lightning bolts can range from the diameter of a pencil to that of a tree trunk— possibly a foot across.)

Ever notice how lightning appears to flash a few times? Lightning does *not* strike as a single stroke from cloud to ground. In fact, chances

41

are the glowing lightning bolt you see is actually traveling from the tree to the cloud and not vice versa!

A faint or invisible negatively charged "leader" leaves the cloud and zigzags its way downward in a series of steps, each about 150 feet long. The highly branched leader "grows" downward like roots of a tree, pausing between steps for just a few millionths of a second. Whichever tip of that root system gets to the ground first determines what gets struck. It's a highly random occurrence.

The last step of the step leader might pause 100 to 150 feet above the ground. At that time the electric field between leader and ground is large enough to make positively charged sparks from one or more objects on the ground near that leader—a tree or a house or a person standing up—jump upward to meet the last step coming down. At the moment of that meeting an electrical circuit is completed. A pencil-thin conducting path is created between the cloud and ground (or you or a tree) and a white-hot flash surges back along the path to the cloud. This *return stroke* is what we call lightning: a colossal flow of positively charged electrical current—forty thousand amperes— from the ground to the negatively charged base of the storm. This phenomenon happens so fast we naturally assume the bolt came from above rather than the other way around. More strokes follow, usually about three or four leader return strokes per flash.

How lightning "picks" its target is a matter of chance. Common belief is that lightning simply strikes the tallest objects. This is because the tallest objects in the immediate vicinity of the leader will initiate the upward discharge that connects first. But object height is not the most important factor; vicinity to the leader is. Once you get just a hundred feet or so away from the leader, it doesn't matter very much how tall you are. The leader won't "see" you. You could have a three-hundred-foot-high tower two hundred feet away from the leader and the leader wouldn't know it's there. The tower would not be struck. But a shorter object within a hundred feet of the leader would be hit. So if you're out playing golf and happen to be the tallest object in the immediate vicinity of a leader, your chances of being struck are better than those of a tree a lot taller than you a hundred yards away.

Hotter Than the Sun

Illustration 16

If you believe that lightning can't strike the same object twice, watch the Empire State Building or the Sears Tower during a thunderstorm and count how many times those skyscrapers are struck.

As for the thunder, you can't have the bang without the flash. Thunder is caused only by lightning and not by any of the numerous fairy tales you may have heard. How? The white-hot lightning bolt heats the air in its path to 50,000 degrees Fahrenheit—five times hotter than the surface of the sun. This hot air expands rapidly in an uncontrolled shock wave, akin to a jet plane breaking the sound barrier. That audio shock is what we call thunder.

Once you become acclimated to the fury of a thunderstorm, it's fun trying to estimate how far away it is. If you figure that the sound of thunder travels at about eleven hundred feet per second, then it will take about five seconds for it to travel a mile (5,280 feet). So the next time you see the flash, count the number of seconds it takes for the thunder sound to arrive. Divide by five and you've got a crude estimate of the storm distance. If the flash is more than fifteen miles away, you may never hear the thunder. This silent, distant phenomenon is what heat lightning is.

Illustration 17

It may be of little comfort, but any flash you can see will not hit you. Watching a lightning bolt strike a tree or a telephone pole close by is one of the most truly amazing and nerve-racking experiences you will ever have. When this happens, it sounds like something has exploded. The flash and the boom occur almost simultaneously, and there's a shower of sparks to rival the Fourth of July.

43

Still a Mystery

Scientists are still trying to understand how the charges in the clouds develop. The most popular theory was proposed by Lord Kelvin over a hundred years ago and is still widely accepted. Kelvin suggested that precipitation in the cloud, falling from top to bottom, brought negative charges with it, leaving the bottom charged negatively and the top charged positively. Other theories suggest that ice and water droplets, colliding in the windswept cloud, or the freezing process itself cause the charge separation. (At ten to twenty thousand feet above the ground—the realm of storm clouds—the atmosphere is below the freezing point of water, even in summer. So ice crystals are constantly forming and growing.)

A newer theory, first proposed about thirty years ago by Bernard Vonnegut and accepted by many lightning researchers, contradicts this idea. Vonnegut's idea involves the existence of a positive "space charge" located just above the ground. Vonnegut believes this positive charge can be drawn up into the cloud by strong updrafts and left at the cloud's top, where it attracts negatively charged particles. Downdrafts carry these particles to the cloud's bottom.

· ·

TRIVIA: One lightning bolt can contain enough electricity to service more than two hundred thousand homes. One measured "megabolt" contained two hundred thousand amperes. Voltage in a thunderbolt can reach more than fifteen million volts. A thunderstorm's electrical rage is equivalent to the continuous expense of a million kilowatts.

Lightning is very beneficial. It is nature's fertilizer. As it rips through the sky, lightning breaks loose the nitrogen in the air. Nitrogen is a key fertilizing ingredient. The fifteen million to twenty million thunderstorms roaring across the earth each year deposit about a hundred million tons of fixed nitrogen on soil and plants as it rains.

· ·

Why Must Rainbows Be Curved?

My heart leaps up when I behold
   A rainbow in the sky:
So was it when my life began;
So it is now I am a man;
So be it when I shall grow old,
   Or let me die!
—Wordsworth, *My Heart Leaps Up*

You might live your entire life without experiencing the magnificence of a rainbow. Rainbows are among life's joyous accidents. They're not events you can really plan on seeing; a person just happens to be at the right place at the right time. The beach is a good spot to observe one, but it's not the only good place. Once while driving through the redwood forests of California, I emerged from the darkness of the towering trees to stumble upon the breathtaking sight of a rainbow. Actually two rainbows, one inside the other. While marveling at my luck, my traveling companion, knowing my penchant for science trivia, asked me why rainbows are curved. Why can't they be endless multicolored bands unfolding like streamers across the sky? And why do the colored lights always begin at the earth, appear to head upward, but then gracefully arch back to the ground? Could it be the pots of gold at each end weighing them down? What an interesting question! Why *must* rainbows be curved?

The answer, of course, involves understanding what rainbows are and how they're formed.

The explanation begins with rain. Rainbows come out only in the rain or right after it rains because raindrops are responsible for producing the colors. As you know, sunlight is composed of all the colors of the rainbow. Isaac Newton pointed out that sunlight streaming through a prism will be broken up into a spectrum of colors ranging from red to violet. A drop of water will have the same effect. As shown in

45

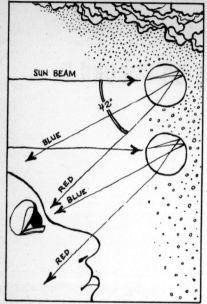

SUN BEAM

42°

BLUE

RED

BLUE

RED

Illustration 18

Notice how each raindrop contributes one particular color to the rainbow. The color contribution depends on the rainbow's height in the sky.

Illustration 18, a rainbow's development begins when sunlight enters the front of a raindrop, gets bent, and is separated into its constituent colors. The colors bounce off the back of the drop and are bent again exiting through the front.

So out of white light comes color. Notice that the light enters the drop in a straight line but leaves the drop having been bent (refracted) and bounced by the water. The amount of bending varies from color to color. So as they come out, the red and blue rays—and all the colors in between—exit at different angles and go in different directions. The angle of deflection for each color also varies, from about 40 degrees for blue to about 42 degrees for red, with gradations of other colors in between. The difference in the angle of deflection explains why the colors are spread out into a multicolored band, red at the top and blue near the bottom.

It would appear that if all it takes to turn sunlight into colors is a collection of raindrops, we should be seeing rainbows quite often—during every rainstorm and shower, in fact. But we don't. Why? It's true that rainbows *are* being created, but to see them, you and the sun must be in the right position with respect to one another. The sun must be at just the right height in the sky so that the angle formed by the sun, the raindrop, and you is just right. This angle is the same angle of deflection shown in the last illustration, about 42 degrees. For the rainbow to be seen, the sun must be positioned in the sky so that the light entering the drops (sunlight) and the colors meeting your eyes form this critical angle.

If the sun is too high—let's say halfway up the sky or at high

46

noon—the rainbow will shoot over our heads and we'll never know it is there. If the sun is too low in the sky at dawn or sunset, the light is driven into the ground before it ever reaches our eyes. But in between these extremes—late in the afternoon or early in the morning—the angle is favorable. The sun is in just the right position; the angle between the incoming light and the observer's eye is 42 degrees, and the rainbow shines in all its colored glory. Notice that the sun must be behind you (Illustration 19) in order for you to see the rainbow, and the light must shine on distant air laden with storm raindrops; it's like being in a movie theater with the sun acting as film projector and the dark storm clouds serving as the screen.

Illustration 19

Despite the fact that each raindrop can produce a whole spectrum of colors, at any given moment each individual raindrop is contributing just one color to the rainbow. Raindrops near the top contribute to the red band; those near the bottom are shining blue, as in Illustration 18. As a drop falls, its angle to your eye changes and so does the color of its light that is reaching your eye. The drop, once shining red at the top of the rainbow, will now shine violet as it falls to the bottom, with a gradation of colors in between. Since no two people can be in exactly the same viewing position, everybody sees a different rainbow. Your rainbow might be coming from raindrops

47

Illustration 20

right next to, or above or below, the ones of your neighbor. Each rainbow is unique and one's very own.

Consider the view if all the colors from each drop could be seen. The sky would be randomly filled with billions of tiny rainbows every time it rained, the rainbows would recombine, and all we would get would be a wall of dull white light. Not very pretty. But thanks to a very considerate Mother Nature and her band of 42 (degrees), we get a neatly organized bright band of color.

Finally, about that question of why rainbows are bowed. Try this thought experiment. Give yourself a paintbrush and canvas (or paper and pencil) and the job of painting a band in the sky, knowing that (1) your hand, representing a person viewing the band, had to remain fixed in position and (2) the band had to be maintained at about a 42-degree angle wherever you looked. How would you draw it? The only way would be to sweep your hand back and forth across the paper without moving your fingers or lifting your hand.

Isn't it intuitively obvious that the only way to maintain the angle is to create an arc? Only in an arc, or in a complete circle, are all the raindrops at the same angle to the sun. Any drops above or below the band would violate the 42-degree angle.

As you walked, the rainbow would "walk" with you, maintaining the same angle. Furthermore, if the ground could somehow be taken away, the rainbow would continue to arc and come around to form a full circle. One way to remove the ground is to get high enough in the air. Rainbow "circles" are quite common sights to occupants of an airplane flying by sunlit clouds, or perhaps to someone looking over the edge of a high, steep mountain, as in Illustration 20.

. . . .

TRIVIA: Double rainbows are not uncommon. Sometimes a second rainbow is seen higher in the sky but with reversed colors. Light entering the raindrop may bounce twice before exiting and be concentrated at an angle of 51 degrees. The area between the two rainbows is markedly darker than the rest of the sky. This section, called "Alexander's dark band," is named after Alexander of Aphrodisias, who first noticed the difference.

. . . .

The Strongest Force on Earth

Oceans and hurricanes are inextricably linked. No hurricane could develop without the ocean, from which it draws immense power.

The most powerful hurricane ever recorded struck Mississippi in 1969. Hurricane Camille packed winds of over two hundred miles per hour. She was so powerful that scientists could only guess what her top wind speeds were because their instruments went off the scale. Steady winds registered 175 miles per hour, with gusts of over two hundred.

As for storm damage, hurricane Frederick holds that record. It laid waste to Mobile, Alabama, in 1979, causing two billion dollars' damage in wrecked homes, cars, buildings, and other property. Although these are the severest kinds of hurricanes, each year "average" hurricanes destroy millions of dollars ' worth of property and threaten the lives of millions of people. Yet like many things in nature, hurricanes are unpredictable and not very well understood by scientists. There is no way to predict where a hurricane will begin or when.

We do know that most hurricanes are born in the hot, wet, tropical regions of the world. The tropics supply the basic ingredients for a hurricane: hot weather, warm ocean water, and a random thunderstorm. But from there it's anyone's guess where they will end up.

A Great Heat Engine

Consider the life of a typical hurricane. It doesn't start out as a hurricane. It begins life as a regular, harmless thunderstorm called a tropical depression—a wave that forms, of all places, off the coast of Africa. From there our thunderstorm begins to drift westward, carried by the winds called the "easterlies" (because they move from east to west). As prevailing winds carry it over tropical waters heated all summer by the sun, heat rising from the ocean surface energizes the storm and makes it grow. Lightning bolts crash. Rain intensifies. Winds begin to howl. The storm becomes larger and more powerful. Water vapor evaporating into the atmosphere causes heavier rains, which release more heat, more energy, leading to a crucial phase: a drop in pressure at the storm center.

No longer is our storm merely a wave of low pressure. Now she is closing ranks. With help from the rotation of the earth, the storm forms a tightly knit circle of wind and clouds. The winds swirl counterclockwise, causing the storm to spin like a top. That's why a hurricane is also called a "tropical cyclone." When its winds reach thirty-nine miles per hour, the rain and high winds are officially classified as a "tropical storm." At a wind speed of seventy-four miles per hour or more the storm is labeled a hurricane.

At the center is the crucial area of low pressure called the "eye." The eye is one of the most fascinating and important parts of a hurricane. The eye is what makes tropical storms and hurricanes different from all others. It is the extreme drop in atmospheric pressure in the eye that makes the storm so powerful. If our hurricane is an average one, it is very large—about three hundred miles in diameter. It is also very powerful. The amount of energy driving a hurricane in one day is estimated to be equal to the daily amount of electricity consumed by everyone and everything in the whole United States.

Don't let the size of a smaller hurricane fool you. Just because it looks small doesn't mean it's not powerful. The tighter the winds circulating around the eye, the stronger those winds are. Like an ice skater who brings in her arms to spin faster, a compact hurricane's winds may be much more fierce.

Storm Surge

It's not the wind, though, that's the most dangerous part of a hurricane. It's the water, especially when something called the "storm surge" occurs. As the low-pressure eye of the hurricane sits over the ocean, the sea level literally rises into a dome of water. For every inch drop in barometric pressure, the ocean rises a foot higher. Now, out at sea, that means nothing. The rise is not even noticeable. But when that mound of water starts moving toward land, the situation becomes crucial. As the water approaches a shallow beach, the dome of water rises. It may rise ten to fifteen feet in an hour and span fifty miles. Like a marine bulldozer, the surge may rise up twenty feet high, crash onto land, and wash everything away. Then with six- to eight-foot waves riding atop this mound of water, the storm surge destroys buildings, trees, cars, and anything else in its path. It's this storm surge that accounts for 90 percent of the deaths during a hurricane.

People in the path of a hurricane may think they can survive the high winds. And in some cases the winds may not be life-threatening. But many people do not realize how powerful the waters of the storm surge can be near the shore. They gamble foolishly and lose their lives in one of Mother Nature's most powerful surprises. Sometimes a storm surge will even arrive a few hours before the hurricane itself, catching people off guard.

Landfall

Once a hurricane hits land, it may begin to subside. First there is the effect caused by the friction of the land. As the hurricane moves inland, the ground, the trees, and the mountains slow the winds down. The hurricane begins to lose steam.

Once on land, the hurricane also loses its source of energy. Without the hot ocean water to supply the hurricane with heat and moisture for rain, the storm cannot replenish itself. So it loses power.

Third, the hurricane may encounter other weather patterns that upset its tightly wound shape. As it heads north, a cold front may bump into it, disturbing the circle of wind or cooling it off; or upper-altitude winds may literally blow the top off the storm. When this happens, the hurricane dies very quickly.

Sometimes a hurricane will move onto the land, lose power, but then change direction, head out to sea, and build itself up again to hurricane force. There is no telling in which direction it will go. Hurricanes follow the prevailing winds, the steering currents. The storm might travel as fast as fifteen or twenty miles per hour. Or it might just loaf along at a slow walk, one or two miles per hour. It might even just sit in one place, spinning like a top for a few hours, not going anywhere. Hurricane Diana did that in 1984.

One thing is certain. Hurricanes occur only during a certain time of the year. The hurricane season stretches from June 1 to November 30, six of the hottest months of the year. But because lots of heat is needed to spawn a hurricane, most hurricanes occur during August— a full 75 percent of them. During these months, one tropical wave after another comes off Africa. That averages out to about sixty storms per year. But only one out of six storms forms into a full-blown tropical hurricane. The rest never achieve the right combination of weather and ocean temperature. The ocean water itself has to be quite hot, at least seventy-five degrees Fahrenheit.

A check of the record books show there were many more hurricanes during certain years than there were in others. Why? Because every year there is a different weather pattern. Where hurricanes are concerned, we're talking about the upper-air circulation pattern in twenty-thousand- to forty-thousand-foot-altitude.

If the prevailing easterly winds are deep and strong, more tropical storms and hurricanes come our way. But if winds from the west are strong, as they were during the years 1982–84, then relatively few hurricanes blow in from the tropics. The westerly winds meet the easterlies head on and shear off the tops of the thunderstorms. That's the end of the hurricane.

With an average of about ten hurricanes and tropical storms per year, scientists would like to find some way of predicting where they will go and how fast they move, but so far that has been impossible to do. Weather prediction is very difficult under the best circumstances, not to mention the worst, and predicting the path of a hurricane is like guessing the direction of a leaf on a stream. It changes from minute to minute.

That doesn't mean that scientists watch helplessly during a storm. Every hurricane is monitored by satellite and airplane. Satellites relay pictures of the hurricane that tell us how big it is, what direction it has come from, and where it might be going. The National Weather Service flies airplanes into the hurricanes, gathering information about wind speed, atmospheric pressure, and motion of the storm.

Now, if you're wondering how it's possible to fly into a hurricane and live to talk about it, consider this: the speed at which an airplane flies versus the wind speed of a hurricane. While a hurricane's winds can peak at one hundred or two hundred miles per hour, an airplane can fly much faster. So another hundred-mile-per-hour wind tacked on to a speedy plane just makes for a faster ride. Since hurricanes are only fifteen to twenty miles wide, the severe wind area lasts for only about two minutes of flying time.

In fact, that's how scientists on board the plane measure the wind speed: They measure the speed of the plane when it's flying through the storm and subtract the speed of the plane when it is out of it. The result is the wind speed of the hurricane. Forecasters on the ground also use radar to measure the wind speed. Ground-based Doppler radar is able to determine if a target in the air (in this case, the wind) is moving toward or away from the radar antenna and at what speed.

A Swirling Doughnut

Once you're at the center of the hurricane, something very strange happens: The winds die down. There, in the eye, the weather is relatively calm and clear. That's because the eye is formed when the rising warm air in the clouds reverses its direction and sinks down into the center. The sinking motion clears out the clouds in the middle, opening up a hole. So a hurricane is like a swirling "doughnut." Inside the doughnut is where the tremendous drop in atmospheric pressure occurs. Normal barometric pressure is about 30.00 inches of mercury. Usually the barometer operates with ups and downs of an inch or an inch and a half. Even the biggest winter storms tend to go no lower than 28.5 inches. But a hurricane can drop the barometer almost 2.5 inches farther—the lowest pressure ever recorded was 26.35 inches. As evidenced by the destructive force of a hurricane, that 4 inches of mercury reflects a tremendous change in the weather, especially in the effect it has on storm surge.

The information collected by airplanes and satellites is fed into a computer. The computer digests the data and gives out its best estimate of the direction the hurricane is expected to take. But the computer forecast is an educated guess at best. The earth's atmosphere is so large that even if there were some way of measuring millions of individual points in the atmosphere, no computer on earth would be big enough to process all those measurements. So there is no way of knowing where a hurricane will be from one hour or minute to the next. Forecasters cannot predict with certainty the direction the hurricane will take. That's why you will never hear a reputable forecaster make a prediction about where a hurricane will be the next hour or the next day. All they can really tell you is where it has been and in what direction is it moving at the moment. That's also why hurricane forecasters will issue a hurricane warning for as wide an area as possible—to allow for a margin of safety.

What should you do, then, if you're caught in the possible path of a hurricane? Here is a list of do's and don'ts as prepared by the National Hurricane Center in Coral Gables, Florida:

1. *Get away from beach areas.* Go inland to avoid the storm surge. Listen to the advice of local authorities, and don't assume you can safely "wait out" the storm in a beach area.
2. *Stay away from windows and the danger of broken glass.* Once inside your home, stay in the interior of your house. Chances are you will be safe even if the roof blows off. Hurricane King in 1950 blew off the roofs of seventy-five homes in Florida. A law in Florida now requires that all new homes have roofs tied down with steel brackets.
3. *Keep an emergency supply of special items: bottled water, canned goods, candles, flashlights, and batteries.* Have a portable radio handy (with fresh batteries) so you can listen to weather advisories should the power go off.

Fortunately, we live in a time when there is plenty of help around in case of a hurricane and plenty of warning should a hurricane cross our paths. That's why some of the worst hurricane disasters occurred before radio or television or satellites were invented. None of today's modern technology could be used to warn people that a hurricane was coming. In 1935 a hurricane swept over Florida, drowning four hundred World War I veterans helping to build a railroad. The storm closed in unexpectedly, packing winds of about two hundred miles per hour. It created a twenty-foot storm surge that left the islands of the Florida Keys under ten to fifteen feet of water.

How the Hurricane Got Its Name

The invention of radio did lots to save lives. As radio came into its own, ships at sea could flash storm warnings as they passed through hurricanes. Small islands in the path of a hurricane could warn others about an approaching hurricane.

As communications expanded worldwide, more and more storms were being reported. And that created a problem. So many people were reporting storm sightings that meteorologists couldn't distinguish one hurricane from another. What to do? Why not give hurricanes names? So starting in 1950, hurricanes were given names taken from the phonetic alphabet. Names like Abel, Baker, Charlie, and David, the kinds of names that radio operators used when talking to one another.

As hurricane forecasting got better and better, other countries joined in. But these countries wanted to use names familiar to them, too. So the names were changed. Hurricanes were given women's names. But that sexist convention didn't last long. In 1977 the World Meteorological Association, which is charged with drawing up the names, included men's names in the list. And now each year, men's and women's names alternate. And in the spirit of international cooperation, you can find German, Spanish, and French names included as well.

Will they ever run out of names? The year that holds the record for the most hurricanes ever recorded occurred is 1933. That year saw twenty-one hurricanes. And since there are twenty-six letters in the alphabet, forecasters are pretty sure they won't run out of names. By the way, the list of names repeats itself every seven years.

· ·

TRIVIA: The worst hurricane disaster occurred in 1900 in Galveston, Texas. Without any warning at all, a storm surge of fifteen to twenty feet swept over five-foot-high Galveston Island and drowned six thousand people. It remains the worst catastrophe ever to hit the United States.

· ·

II. Some Notes Are Made to Be Flat

You need three things in the theater—
the play, the actors, and the audience,
and each must give something.
—Kenneth Haigh

The Best Seats in the House

Going to the theater these days is not an inexpensive proposition. A ticket can run upward of fifty dollars or more. Therefore, that hundred-dollar bill should buy a decent pair of seats at the theater, concert hall, or opera house. Those seats should be so good that no matter what their location, the music and dialogue onstage come through loud and clear. How can that be assured?

Music halls, when designed carefully, are marvels of modern acoustical engineering. You just can't place an orchestra or opera company in a box and expect it to sound like a stereo recording. The acoustics have to be painstakingly planned to the last detail. Most of the world's great houses of music are crammed with gimmicks and gadgets you'd never guess were there to make the music sound better. Even the chairs are special. And the chandeliers as well. I'm going to use the John F. Kennedy Center for the Performing Arts in Washington, D.C., as an example of a well-designed modern structure. You can compare notes the next time you're there.

Whether you find yourself in the Eisenhower Theater, the Opera House, or the Concert Hall, look around at the general shape of the room, the fixtures, the carpets, the drapes, and the upholstered seats. They have all been meticulously tailored to control one major factor: the reverberation of sound in the room, that is, how long sound takes to die down and how smoothly.

The nature of reverberation is that sound persists after the source has stopped making the sound—this is what we call an echo. If that sound is strong enough, the reverberations may last for a long time and the room is described as being "live." On the other hand, if the reverberations are absorbed, reverberation time is short, and the room is described as being "dead." For most purposes people prefer a live room versus a dead one; the sound has a fullness of tone. But given the wrong conditions, a live room can be quite annoying (see Illustration

59

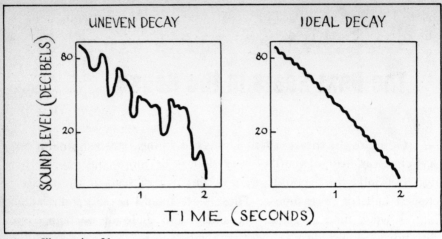

UNEVEN DECAY   IDEAL DECAY

SOUND LEVEL (DECIBELS)

TIME (SECONDS)

Illustration 21

The graph on the right shows ideal conditions, the sound in a concert hall dies away smoothly in about two seconds. The graph on the left shows concertgoers' nightmare: The sound decay is erratic. The music sounds terrible.

21). The reverberations may mask any new sound that starts before the first has stopped. Speech, for example, is a series of short syllables. If the reverberation time of a room is high, then the reverberant sound energy of the first syllable may mask the sound of the next syllable as it leaves the mouth of the speaker. That's why it becomes so hard to understand the person next to you at an indoor swimming pool, or a speaker in a very large auditorium: The reverberation time is so long that his last words are still bouncing around the walls and ceiling, masking the new words coming out of his mouth. The words become garbled.

The same theory is true for musical notes. Uncontrolled reverberation of the music around the room can result in notes masking or drowning one another out. Quiet notes may never be heard because they have little chance of surviving among their louder, more reverberant brothers.

Performers as well as listeners need to feel comfortable. With too little reverberation, performers feel like they must struggle to fill the room, making them feel frustrated or fatigued.

The art and science of music hall and theater design, then, is massaging the reverberation: designing the hall or theater to meet the kind of music or drama that will play there. Since music needs more reverberation than speech, a room made for music will be designed differently from one made for drama. For example, longer reverberation time, desirable in a very large room like the Concert Hall, may be overpowering in a smaller room like the Eisenhower Theater. So rooms used for piano recitals and chamber music need short reverberation times (typically under one second), larger halls used for full orchestras have longer reverberation times (one to two seconds), and halls that house choral music sound best with the longest reverberation times (up to 2.5 seconds—the reverberation time for the Thomaskirche in Leipzig). Of course, all of these designs assume a full house, but more on that later.

So much for theory. In actual practice how does one control reverberation? In lots of different ways. First, let us notice the seating and accessories in any of the rooms. The plush chairs, carpeting, and curtains may be beautiful to look at and wonderful to feel, but clever architects have placed them in the room to absorb sound and soften reverberation.

The shape of the room also contributes heavily to the effects of reverberation. Looking around any room again, what don't you see? A rectangular box of a room with bare walls and ceiling. It's not a good idea to design a room composed of an even number of flat reflecting walls and a flat ceiling. Because then the sound is allowed to reverberate at will like a ball in a racketball court. The ball hits some parts of all walls and the ceiling but misses most. If that should happen in a concert hall or an opera house, some parts of the room may be alive, while other parts may be quite dead. Some people may hear the strings while others get only a full blast of the horns. Everyone will want their money back.

What can be done? The idea is to scatter the sound evenly in every direction so that each seat is a good one. Whether you're sitting in front-row orchestra or last-row balcony, the sound should be equally distributed. A good design will use every trade secret to diffuse the sound evenly, and the first secret is to make the room irregularly shaped.

Concert halls of the 1800s (like Carnegie Hall) were generally rectangular, but they had one thing going for them: ornate decorations on the walls, which diffuse the sound in all directions. Cornices, friezes, cherubs, busts, and extravagant ornamental plasterwork may have looked gaudy, but they distributed the sound very efficiently. Symphony Hall in Boston was the first concert hall in the world designed by the father of modern acoustic design—Professor Wallace Clement Sabine, a physicist at Harvard. Insisting they were necessary to diffuse sound, Sabine placed eighteen replicas of Greek and Roman statues in niches around the walls above the second balcony. Although the statues did not meet the tastes of the Bostonians of 1900, they did the job, and the acoustics of Symphony Hall are commonly agreed to be excellent, making it one of the top-ranking auditoriums in the world for concert music. (Incidentally, its reverberation time is slightly over one second.)

Illustration 22

Modern concert halls achieve the same effect less dramatically by building shapes and patterns into the walls, tiers, and ceilings. For example, stroll into the Kennedy Center Opera House and look at the box tier. The boxes provide more than just a nice place to drink champagne. The faces of the boxes are bowed out into convex panels, which diffuse the sound that strikes them. Boxes break up the smooth lines of the walls of the Royal Festival Hall in London and at other major opera and concert halls around the world.

Observe the walls. If you find an odd-shaped "whatsit" that looks like a work of modern art but appears to serve no conspicuous purpose, chances are it's there for the acoustics. The back and side walls of the Opera House are formed by a row of convex cylindrical surfaces running from floor to ceiling. These are not an interior designer's experiment in modernistic columns but a tried and true form of sound diffusion. Convex surfaces, be they fascias of a box or cylinders of a column, scatter the sound in all directions.

The Concert Hall diffuses sound by another technique: rough surfaces diffuse sound very well, so the walls are given a sawtooth rake. A series of panels are positioned in front of the flat walls, leaving an air gap between the walls and panels. Walls with recesses and air gaps look very rough to sound waves. To fill the performers' need for additional reverberation, the stage is surrounded with hard reflecting surfaces, which also help to direct the sound out to the audience.

Now glance at the ceiling. The ceiling might be made up of various-sized hexagons, as it is in the Concert Hall, or a series of concentric, stepped circles, as in the Opera House.

Even the chandeliers serve more of a purpose than meets the eye. The Concert Hall is festooned with eleven beautiful crystal chandeliers, gifts of the government of Norway, each weighing a hefty twenty-seven hundred pounds apiece. The crystals are circular and arranged in closely packed hexagons, forming three tiers. This dazzling design was no accident. The glistening, delicate crystals are very good at diffusing sound in several different frequency ranges.

If a note is to be diffused efficiently, the size of the diffusing obstacle must match the note's wavelength. A middle C tone is four feet long and would ignore a one-foot diffusing ornament. A four-foot niche in the wall would diffuse middle C but would hardly affect the C two octaves below, which has a sixteen-foot wavelength. High-frequency tones (short wavelengths) are easier to diffuse and can be scattered by busy ornamental plaster or ornate crystal. So a good concert hall will have diffusing obstacles of all lengths and sizes to scatter both high and low notes efficiently.

No object in the hall can be overlooked. The chairs have been carefully selected for their sound-absorptive qualities. So has the carpet, down to the composition and depth of its pile. At the Kennedy Center the carpeting used in the interior of all three main auditoriums was chosen *not* to absorb sound—a blend of 70 percent wool and 30 percent nylon, with a short pile. By contrast, in areas outside the rooms the carpeting does absorb sound. It is all wool, and very heavy, with a deep pile.

Most of the great opera houses of the world are horseshoe-shaped. Why? Opera houses are usually larger than theaters, and their size puts a great strain on a singer's voice. A horseshoe shape provides the short sound path needed between singers and their audience. On the other hand, when there is fast dialogue being spoken, as in a play, an opera house is too big. That's why architects design theaters to be small and intimate; then the entire audience, and not only those in the front rows, can hear even the stage whispers.

Architects could spend their whole lives designing perfect concert halls, opera houses, and theaters and it would all be for nothing if outside noises interfered. Ambulances, trucks, airplanes, and productions in adjacent theaters in the building can cause noise to leak in and ruin a performance. What can be done? The answer is to build the rooms as a box-within-a-box. At the Kennedy Center, each room is isolated from the rest of the building by another room that surrounds it. Walls may also be double-walled, filled with insulating fiberglass blankets. The designers of the Kennedy Center lined the walls in each room with a one-inch layer of cork.

The next time you curse those thick, heavy doors that have to be wrestled open just to get outside the room, be thankful instead. Those "sound locks" are there to make your evening more enjoyable.

There's one final design characteristic of note: you. The audience affects the acoustics. Heads are good diffusers of sound, so halls are designed for capacity audiences. Unlike a chandelier or a carpet, a packed house is the one intangible architects include in their designs but can never guarantee the delivery of.

TRIVIA: Two days before the official opening of the Eisenhower Theater of the Kennedy Center, Sargent Shriver fired a small cannon in the half-full theater. A cannon is used by acoustics experts to record and analyze the reverberation time of sound in the theater.

In an ideal auditorium, the sound traveling directly from the performer and the sound reflecting off the walls and ceilings should arrive at the listener's ear no more than one twentieth of a second apart. If this occurs the reflected sound will appear to reinforce the direct sounds, and the result will be appealing. If the time difference is greater, the resulting sound will be perceived as an annoying echo.

The Right Tool for the Right Tone

Had I learned to fiddle, I should have
done nothing else.
—Samuel Johnson

Why don't all the instruments in the orchestra sound alike? If they are all playing the same notes, why doesn't an oboe sound like a flute or a violin like a trumpet? Why do the timpani boom out a middle C while the oboe sounds it quietly? After all, shouldn't an A sound the same no matter what instrument plays it? But the fact is, it doesn't.

On the other hand, what could be more nonsensical than to compare brass instruments with strings? On one instrument, the musician blows intensely through a small hole in a piece of twisted metal. On the other, a bow is dragged across tightly stretched strings. How *could* they both sound alike?

Of course, the material of which the instrument is made makes a great difference in the sounds of the notes. Also, the method of making the sounds—blowing, striking, plucking, and bowing—also affects the nature of the sound. But that explains how the different notes are made. It doesn't explain why they sound different!

The answer to this question is elegant and, literally, easy to see. The sound waves produced by different instruments *look* remarkably different. Look at middle C produced by an electronic soundmaker, as in Illustration 23, top. It's a pure note that looks like a wave on a tranquil sea. Now look at middle C made by striking the key on a piano as in Illustration 23, bottom. What's happened?

Our peaceful little note now has little waves riding on top of it. It is no longer pure; there are other notes mixed in. Why does this happen? Because the piano string is not sitting alone. Surrounding it are the other strings and the sounding board, which vibrate and add sounds of their own to the original note.

66

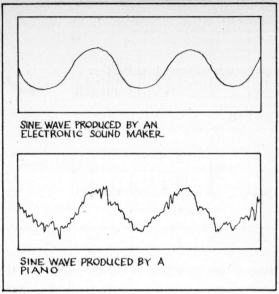

SINE WAVE PRODUCED BY AN
ELECTRONIC SOUND MAKER

SINE WAVE PRODUCED BY A
PIANO

Illustration 23
Sine wave

These other notes given off by the piano are not as strong as the "fundamental" note struck by the hammer. But these "partials," "harmonics," or "overtones" are strong enough to be heard. It's the harmonics of each instrument that give each instrument a sound like no other.

Here, in Illustration 24, is an acoustic spectrum of two instruments, showing their fundamentals and harmonics. The strength of each harmonic is shown as a vertical line. The pitch (frequency) of the fundamental note, in this case A above middle C for piano and violin, is 440 vibrations per second (Hz). Each harmonic is one octave above the other. Harmonic 1 is also the fundamental.

The figures are surprising. The fundamental is the note that we play and expect to hear. But you can see from the illustrations that the fundamental of a piano contains only a fraction (32 percent) of the sound energy of what we actually hear. The other harmonics (H2-H10 and higher) contain the rest of the sound. However, since the second strongest overtone—harmonic 2—is just one octave higher, the piano still sounds harmonious. You can hear that for yourself when you play two notes an octave apart.

67

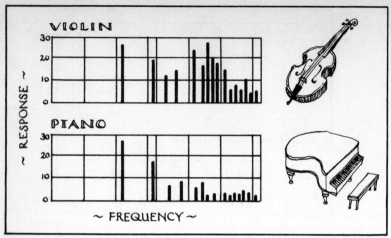

Illustration 24

Now compare this to the same A note played on the violin. The fundamental of violin is still the single loudest sound. But look at harmonic 7: It's just about as loud as the fundamental! And harmonics 5 to 9 together contribute much more sound than the fundamental. These strong harmonics are of higher pitch; that's why a violin sounds "squeakier" (heaven forgive me) than a piano—a violin's sound is produced mainly in the higher partials.

Let's compare two "human" instruments: a soprano and a bass singing the vowel sound "ah," as in Illustration 25. Look at the acoustic spectrum:

The soprano's harmonics are bunched like a picket fence in the higher registers, while the strongest bass sounds are concentrated in the low harmonics. Is there any clearer proof why a soprano and a bass voice sound so different?

No instrument plays a purer note than the flute, and a look at the spectrum shows why. Here, in Illustration 26, are the acoustic charts of a flute playing the note G one and two octaves above middle C.

The higher note produces an almost pure tone, with just one harmonic that barely exists. Even the lower note's fundamental carries the great bulk of the sound.

It's becoming obvious why instruments don't sound alike. They have different harmonics, which vary in degree from one kind of

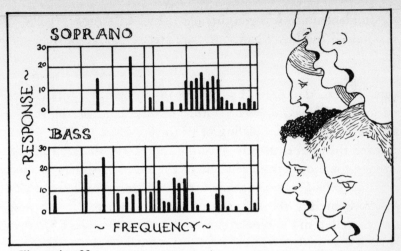

Illustration 25

The acoustic spectrums of soprano and bass voices producing the vowel sound "ah."

instrument to another. The flute (and its cousin the organ) produce notes that come out strong and pure since the higher harmonics (those far from the fundamental) are very weak. The low notes on the cello sound full and mellow because of strong lower harmonics (those closer

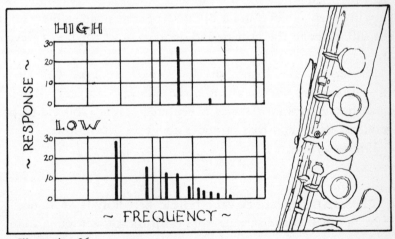

Illustration 26

The acoustic spectrums of two notes of a flute show very few harmonics, yielding clear and clean notes.

to the fundamental). If the higher harmonics are strong, we get shrill, thin notes. If the lower harmonics are strong, we get a rich, mellow tone.

What makes some harmonics weak and others strong? Many factors. The wood or brass of which the instrument is composed, or the vibration of the reed or reeds. Other factors could be the shape of a trumpeter's lips or the opening of his or her horn.

Even the lacquer used in finishing the wood will determine which overtones come out stronger. One of the great mysteries in music is why violins made by Antonio Stradivari are considered to be so superior. Some say the secret is in the varnish used in treating the wood, as well as in the type of construction, which gives a Stradivarius its inimitable quality.

But even if all the harmonics could be stripped from the different instruments and the notes compared, there would still be notable but slight differences due to variations in growth and decay of the sound. A slow growth of sound will yield a different tone than a rapid growth. A piano string, for example, is struck by a hammer so the sound is immediately loud and decays gradually. But consider the harpsichord. Its strings are arranged similarly, but the sound is obviously different because harpsichord strings are plucked, not struck. The difference in quality between an oboe and a B-flat saxophone is due almost entirely to the difference in the growth and decay time of the notes played by each.

In humans the shapes of the sound passages—nose, throat, larynx—influences the sound and make each of our voices unique. The band of frequencies for females is centered around 260 Hz (approximately middle C) and that of males about an octave lower.

Vive la différence!

. . . .

TRIVIA: The human ear can hear incredibly faint sounds. In normal conversation, the eardrum moves only four thousandths of a millionth of an inch. The membrane of the middle ear, called the *oval window,* moves even less. The weakest sounds humans can hear move the oval window only three hundred thousandths of a millionth of an inch—

equivalent in movement to the thickness of a sheet of paper compared to the distance between New York and London.

Pitch is a physiological phenomenon and is perceived differently by each of us. In general, as the frequency of a sound increases, so does the pitch. However, pitch also depends on loudness. If the loudness of a 200 Hz tone is increased, the pitch decreases, while for tones above 4000 Hz, turning up the volume lowers pitch.

. . . .

III. Newton's Backhand: Sports Secrets

. . . .

When the One Great Scorer comes
　　to write against your name
He marks—not that you won or lost
　　—but how you played the game.
—Grantland Rice

The world of sports is filled with mythology and superstition. Baseball players religiously avoid washing lucky shirts that bring hits. Golfers use lucky long-driving tees of one color. But successful sportsmen know as much about the science of their sport as they do about its superstitions, and they make good use of that knowledge. It goes without saying that intuition plays a large part in knowing the "right" way to hit a ball or how to execute a tricky slalom turn, but when analyzed, each intuitive feeling has a sound basis in science.

Illustration 27

What's the Racket?

There's no use fooling ourselves, we'll never play tennis as well as McEnroe or Navratilova, even on their bad days. But we can get an edge on weekend opponents by understanding the hidden secrets of the tennis racket.

I say secrets because a lot of tennis snobs think they've learned all there is to know about tennis rackets by hanging out at the local pro shop. What the pros never tell them (because perhaps even they don't know) is that the tennis racket is a complicated piece of equipment. A lot happens when ball meets wood, metal, and gut.

Take the sweet spots, for example—that's "spots" with an "s." Most people think that the tennis racket has just one sweet spot: the place on the string that gives you the most for your swing.

For example, many players know that the spot where the ball hits the racket can make a big difference in the outcome of their swing. After hitting a poor shot, a player may say the ball hit strings too close to the frame or the racket turned in his or her hand. Or, for lack of a precise reason, usually the "it just didn't feel right" explanation will suffice. But there *is* a spot on the racket, most agree, that produces a solid return and a satisfying feeling of striking the ball "just right." Tennis players call this site the "sweet spot": the place on the strings that yields the most for the swings. Many weekend players will spend countless hours trying to master the art of hitting the sweet spot. But what most of them don't know, no matter what their level of play, is that a tennis racket has more than one sweet spot.

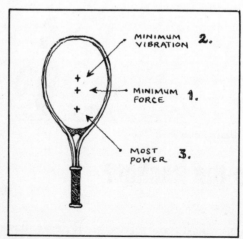

Illustration 28

Sweet Spot 1: Hand Control

When a tennis ball hits the racket, it will force the racket to rebound backward and rotate forward. Both forces try to rip the racket from your hand. But it is possible to have the backward rotation and the forward motion cancel each other out, leaving no force on your hand. How? By always making the ball hit the center of the racket. Try it yourself. Bounce a ball off the strings near the top of the racket and feel how much the racket twists in your hand. Try it again near the center and notice the difference. This sweet spot is the place where the ball is less likely to make the racket come out of your hand.

Sweet Spot 2: No Vibration

When the ball hits the racket, it tends to make the racket "ring" like a bell, that is, vibrate wildly. The racket likes to wag its tail (handle) and nod its head around a central point. But there is a spot near the middle of the racket that relaxes while the rest of the racket does the twist. You might call it the pivot point. Scientists call it the node. Regardless of what you call it, it's the quiet place around which both ends oscillate. And guess what? This point is the place where contact with the ball doesn't create any jarring, annoying gyration. You can find sweet spot 2 right *above* the center of the racket. Hit the ball here and the racket swings smoothly.

Sweet Spot 3: Heavy Hitters

The "power spot" is the place where maximum power is transferred from the racket—and your body—to the ball. This one is easy to find. Lay a racket on a table so the handle is anchored and the strings hang over the edge. Drop a tennis ball onto the strings. The place where the ball bounces the highest is the power spot. Try the top first, and work your way to the bottom. The results are surprising.

Intuition says the ball should bounce highest at the top of the racket. Your mind wants to compare it to a baseball hitting a bat: The fat part of the bat is the spot from which most home runs are hit, so why shouldn't the top of the tennis racket deliver the same amount of power? But in a tennis racket, the power spot is at the *bottom* of the

77

strings, right above the throat. That's where the racket is the stiffest. Here's where more of the muscle you put into the swing can be transferred to the ball. The lower you hit the ball, the more power you have. This is why tennis racket heads are bigger than they used to be: By enlarging the stringed area toward the hands, the power sweet spot is enlarged.

It's also why tennis manufacturers keep coming up with new structural material like graphite and boron. These materials make for a very stiff racket. And that's why tennis mavens go into the pro shops and have their rackets strung very tightly; they know they're going to get the *most* power out of a tightly strung racket. Unfortunately, they're wrong.

When the ball hits the racket, it deforms—compresses and changes shape. In changing shape, the ball loses a lot of energy. The strings, though, are designed not to lose energy when they deform. Tennis racket designers know that if they can make the strings change shape while the ball stays round, the racket will "trampoline" the ball back at high speed and not lose much energy. All the energy stored in the strings will be transferred to the ball.

How can this be done? By not tightening up the strings; loose strings don't squash the ball as much and make a better trampoline.

But why do the *pros* do it!? They really tighten up the tension! Yes, but they do it for a different reason. They're not looking for additional power. Their well-developed musculature supplies all the force they need. The professionals crank up the catgut because they believe it gives them more control and accuracy.

. . . .

TRIVIA: You can enlarge the sweet spots by using an oversized racket. The head of an oversized racket has been made larger by expanding the string area down toward the handle. Therefore, the sweet spot that gives minimum force on your hand is larger. Also, since an oversized racket has more stringing on the lower end, where the racket is stiffest, there is a larger power sweet spot.

A tennis ball spends only about four one-thousandths of a second in contact with the racket. The racket, when it's hit, takes fifteen one-thousandths of a second to come back. So by the time the racket bounces back, the ball is gone. The motion is wasted. That's why most energy is lost at the tip, where the racket is most flexible.

. . . .

A Slice of Life: Why a Golf Ball Has Dimples

If you watch a game, it's fun.
If you play it, it's recreation.
If you work at it, it's golf.
—Bob Hope

The first golf balls were made of boxwood. Whack 'em with a golf club and they made a nice sound, but they didn't go very far. Seventeenth-century technology came up with the featherie, the forerunner of the modern golf ball. The featherie was little more than a tiny version of a crude baseball. The shell was made of cowhide sewn together and turned inside out to hide the stitching. It was stuffed with boiled feathers, stitched shut and left to dry rock hard. The ball was pounded with a mallet to make it round and then painted white. It was a very expensive way to make a golf ball, but this ball carried much farther down the fairway.

Getting the ball to fly far is still the dream and pursuit of every golfer and maker of golf balls. And that's where dimples come in. Golf ball folklore says that somewhere in the mid-nineteenth century, it was discovered that when balls were nicked, cut, or roughened, they flew farther. So why not roughen the surface on purpose and get those extra yards? The first dimples were put on golf balls about 1908 by the Spalding company of America, and golfers have been trying to get the most out of each dimple ever since.

Getting Distance

The key to distance is getting the golf ball to spin correctly. Golfers always try to impart backspin on a golf ball when it's hit. They know instinctively that backspin allows the ball to stay in the air longer. And there is good reason for this.

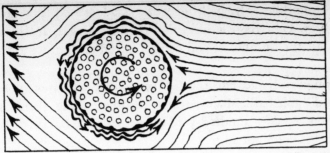

Illustration 29

The dimples of a spinning golf ball scoop up air.

A rough, dimpled surface drags a layer of air with it as it spins (Illustration 29). Since the ball is spinning backward, the air trapped on the top of the ball is moving in the same direction as the air rushing past it as it flies. The air trapped on the bottom is moving against the wind. So the bottom air, which fights the wind, is moving slower than the top air. This means, according to a principle discovered by Daniel Bernoulli, that there is less air pressure on the top of the ball, so the ball stays aloft much like an airplane.

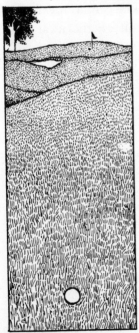

Illustration 30

The aeronautical engineer in you may say, "Why not just put in deeper dimples and make the ball fly forever?" Good idea, but in reality the deeper the dimples, the greater the drag: The wind will slow it down too much. So it's a trade-off: just enough dimpling for distance, but not enough to create too much wind drag.

Topspin—making the ball spin in the direction of flight—is a golfer's nightmare. Topspin gives the ball a flight path the shape of a banana. After a very short ride, the ball dives sharply into the ground. Slices and hooks are just balls spinning right and left, which end up the same way.

In 1975, a physicist and chemist combined talents to produce a weirdly dimpled golf ball designed to minimize hooks and slices. Dimples cover about half its surface, and are deeper on the top and bottom than around the middle. The ball did work as advertised, but needless to say, the U.S. Golf Association ruled the ball illegal. It did not meet the rules for the number of and configuration of dimples allowed in a golf ball.

If you're really adventurous or like to play on a short golf course, try a "bald" golf ball. Made without dimples, these giant "marbles" will shorten any two-hundred-yard drive to a mere hundred. But what will it do for your putting?

· ·

TRIVIA: A drive hit without backspin will come down after four seconds from a height of sixty-five feet. The same drive hit with a dimpled golf ball will stay up for six seconds, resulting in a drive up to eighty feet longer.

Golf became so popular in the fourteenth century that archers put down their bows and picked up golf clubs instead. Since archery was in the national interest, golf was outlawed by the king. Golfers are indebted to a bishop of St. Andrews, who in the fifteenth century, in a royal charter, allowed the inhabitants to play golf on the local links.

· ·

The Most Difficult Act in All of Sports

*. . . the 1962 world championship was finally
determined by an otherwise perfect swing of a bat which
came to the collision 1 mm too high to effect the transfer
of title.*
—Dr. Paul Kirkpatrick, "Batting the Ball," *Am. J. of Physics,* 1963

Hitting a baseball has been described as the single most difficult feat in sports. And for good reason. Imagine the quality of hand-eye coordination required to make contact with a little white sphere traveling at over ninety-five miles per hour, using a 2¾-inch-wide piece of wood being swung at over sixty miles per hour. Consider the intense concentration. A batter standing just fifty-six feet from the pitcher's hand has only about forty-five one-hundredths of a second to decide if he'll swing, predict where the ball will be, instruct his muscles to move, and bring the bat to a point of impact. If all goes well, the bat and ball rendezvous a few inches in front of the plate. The ball is crushed to half its diameter, springs back, and is launched on its return flight at speeds close to a hundred miles per hour. Timing is essential. The difference between a hit over second base and a foul near first or third is a swing mistimed by 0.01 second. Baseball is the only sport where being a failure seven out of ten times is considered to be outstanding—only about a dozen players in each major league bat .300 annually. A basketball center who sank only 30 percent of his baskets or a quarterback who hit his receivers only 30 percent of the time would be selling insurance instead.

Bat Speed
With these kinds of odds working against a hitter, it's no wonder batters will try anything to increase their chances of making solid

Illustration 31

contact with the ball. Some pray. Others choke up on the bat. But no matter what the technique, the aim is the same: bat speed and control. Bat speed is to a hitter what hamburgers are to McDonald's; a batter lives or dies by how fast he can whip that bat around. A hitter knows that to make contact with a blazing fast ball or to follow a sweeping curve, he has to be able to move his bat quickly. For wood to meet leather, a hitter must be able to accelerate his bat from its parked position near his ear and drop it quickly down to the strike zone near his waist, all in just a fraction of a second. And the bat must be moving fast enough at the moment of impact to drive the ball out of the infield or out of the ball park. Bat speed is the essential, because as any physicist will tell you, the faster the bat is moving, the more energy is imparted to the ball. Old-timers used to swing very heavy bats, thinking that by swinging a heavy club they could knock the ball farther. Babe Ruth's bat weighed a hefty forty-two ounces. He once used a fifty-two-ounce bat. He was able to whip his bat around because of his exceptional strength. But in the past thirty years, ball players have discovered that bat weight is not as important as bat speed. A

84

medium-size bat is already six times the weight of the ball. Making it seven times heavier will hardly influence how much momentum is transferred to the ball. But it will slow down your swing considerably.

How is bat speed increased? Watch the lead-off hitters, the small guys who must get on base so the power hitters can drive them in. Lead-off hitters need to be able to bunt, punch the ball to the opposite field, or find the "holes" in the infield, anything to yield a hit. That means they need excellent bat control, and good bat speed. These table-setters are very fast but not usually very big. How do they compensate? By "choking up" on the bat—sliding their hands higher up on the handle. When the thick end of the bat is brought closer to the body, the bat becomes easier to swing. Great hitters know, intuitively, that by choking up on the bat they can swing faster and hit the fast ball they might otherwise foul off. Not many of them know the principles of physics that explain why their techniques are successful.

Swinging a Bat Like a Sledgehammer

The principles used by a baseball player are the same ones a woodsman uses to swing an ax or a sledge hammer. The useful weight of a bat, like the head of a hammer, ax or golf club, is concentrated in one spot: at the point of impact. The farther your hands are from that point, the more difficult it is to lift and control the weight. Your wrists may wiggle just a few inches, but the weighted end of the bat or club will waggle many feet in response. In effect, the shaft is really a long lever magnifying your hand movements and exaggerating the weighted end of the rod. That's why it's much easier to pick up a sledgehammer by its head than to grab it by the end of the handle. Golfers get lots more control from short irons than from long, driving clubs. The same is true of a bat. Shorten the effective length by choking up and you get better control of the weighted business end.

But there's a problem. A short bat is not long enough to reach those fast balls on the outside of the plate. A more permanent solution is to use a lighter bat. When he broke Babe Ruth's lifetime home run record, Hank Aaron's Louisville Slugger weighed only thirty-two ounces. Roger Maris favored a light bat in hitting his historic sixty-first homer. Today's young players moving up through the ranks in high school and the minor leagues use bats as light as twenty-eight ounces. What makes a bat light? Less wood. Since most of the wood is concentrated on the fat end of the bat, most bats are lightened by thinning the handles and hollowing out the ends. These days, unfortunately, lots of ballplayers are breaking their skinny-handled bats and blaming the bat companies for defective lumber. These hitters claim that bats are being made with younger, inferior wood, or from trees that have grown too quickly. But batmakers say the wood hasn't changed. It's still the same northern ash used 102 years ago. If the bat handles are getting "sawed off" in players' hands or shattering into splinters, it's because players are ordering bats too thin to withstand the impact of a ninety-mile-per-hour fast ball.

The Five-Hundred-Dollar Bat

What can be done to solve this problem? The answer is to make the bats out of something other than wood, a lightweight material like aluminum. Today aluminum bats are standard equipment in every ball club *except* those in the major leagues. The official explanation is that major-league baseball fears for the life of the pitcher. As it stands now, a pitcher has just enough reaction time to get out of the way of a wicked line drive or to put up his glove to protect himself. But a lighter, faster aluminum bat would increase the speed of the ball enough to overcome that margin of safety. Some pitchers might get killed or have their careers ended if hit in the head or the elbow. Willie Stargell, the great Pirates hitter, was afraid of killing the fans in the stands who wouldn't have enough time to evade a line drive fouled off an aluminum bat.

Of course, the unofficial reason is that an aluminum bat would upset all the baseball statistics accumulated over the past hundred years, making objective comparisons between players' performances impossible.

Major leagues aside, the wonderful world of bats is not limited to just wood and metal. New space age metals and composite plastics are finding their way into baseball bats. Thousands of amateur baseball and softball players are now using a bat made of graphite, glass, and plastic, the same stuff used to make high-performance airplane wings. Made by the Worth Bat Company in Tullahoma, Tennessee, the composite bats are lighter and stronger than wood, sound more like wood than metal on impact (no aluminum "ping"), and have a "sweet spot" two or three times larger than a wood bat has. The bats are pricey, costing fifty-five to ninety dollars, but they may last a lifetime. Worth's Harold Becklin claims the bats, under normal use, are almost indestructible. But that's not all. The hitters of tomorrow will be able to choose their bats to fit the hitting situation much like golfers choose their clubs for each shot. Composite bats can be tailor-made to enlarge or deaden the sweet spot. Some bats, like golf woods, will be tuned to hit the long ball. Others can be designed like short golf irons for cheap hits that trickle past scampering infielders.

While searching for the perfect bat material, engineers have experimented with everything from Fiberglas to bamboo to magnesium. But there are limits inherent in each material. The latest fiber-imbedded, space age metals, for example, cost upward of three hundred dollars per pound. With each bat requiring a pound and a half of raw material for construction, the price of these bats (not including manufacturing costs) is beyond even the most diehard Sunday softballer.

Center of Percussion

There is something that can be done to make the best use of the bat you have now, and that is to find the best place on the bat to hit the ball, the sweet spot. There are two sweet spots, possibly three, depending upon whom you ask. The first sweet spot is called the center of percussion. It is the place on the bat where the initial jar to your hands is at a minimum when you hit the ball. It's the place where you feel a "solid" hit has been made. If a ball strikes a bat above or below this point, the bat will try to swing around it. The result: The impact of the ball will try to rip the bat from your hands. The center of percussion is not ingrained in the wood (or aluminum) of the bat but moves around depending on where you fix your hands. For an aluminum softball bat about thirty-two inches long, the center of percussion is about 6.5 inches from the fat end of the bat.

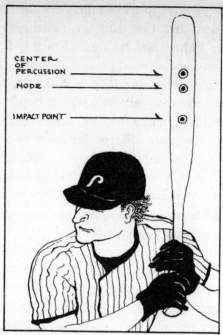

CENTER
OF
PERCUSSION ————————————————→

NODE ————————————————→

IMPACT POINT ————————————————→

Illustration 32
A baseball bat has more than one "sweet spot."

No Sting

Ever wonder why your hands "sting" when you hit the ball? That's because you've hit the ball on the wrong place on the bat. The second sweet spot is the point on the bat where your hands sting the least when bat meets ball. Not to be confused with the center of percussion, the "node" is the spot where a hit will cause no lasting vibration. This point is about a quarter of the length from the fat end of the bat. To find the node, hold the bat by two fingers about six inches from the knob and hit the bat at various points (see Illustration 33).

The bat will ring at each point of impact until you hit the node. The farther away from the node that you hit the bat, the louder the ringing will be. Aluminum bats ring a great deal more; that's why they sting your hands more often.

Fastest Rebound Speed

The existence of a third sweet spot is being hotly debated by sports physicists. This is the point on the bat that, when struck, will

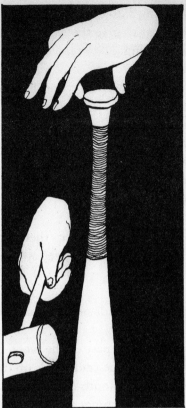

Illustration 33
Finding the node: Person holding bat dangling from two
fingers and hitting it with mallet.

transfer most of the energy from the bat to the ball—in other words,
the spot where the speed of the rebounding baseball is at a maximum.
Many sports physicists, such as Dr. Larry Noble of Kansas State
University in Manhattan, Kansas, place that point at the center of
percussion. Dr. Howard Brody, a physicist at the University of Pennsylvania in Philadelphia, disagrees. Brody says this sweet spot—the
impact point—is not at the center of percussion but at a spot closer to
the handle than the other sweet spots on the bat. This location,
however, moves around depending on the speed of the thrown ball,
the weight of the bat, and how "wristy" the swing is—that is,
whether the arms and shoulders or the wrists of the player are more

89

involved. Without getting into complicated mathematics to prove his point, Brody suggests that to give the ball maximum rebound velocity you should try the following:

1. Hit a *fast* pitch closer to your hands for maximum power. Hitting a slow pitch farther out on your bat gets best results.
2. When playing hardball, where the weight of the bat is much greater in proportion to the weight of the ball, hit the ball farther out on the bat. In softball, hit the ball closer to your hands.

Sunday softballers can decide for themselves from their own experience whether or not the third sweet spot exists.

· ·

TRIVIA: A baseball spends one-thousandth of a second in contact with the bat.

To get the strongest part of the wood on the ball, the bat should be held with the trademark up (toward the sky). The trademark is branded onto the flat growth rings. Since the growth rings are the strongest part of the wood, it's best for the ball to hit the rings on edge. But Hall of Fame catcher Yogi Berra refused to turn the label up. Saying "I don't come up to read but to hit," Yogi always turned his label to the pitcher. Yogi broke lots of bats.

· ·

Debate About the Ball: Does It Really Curve?

A fast ball is not the most difficult pitch to hit. Lots of players make a living hitting good, major-league fast balls. The pitches that give most batters a problem are breaking pitches: curves, sliders, and split-fingered fast balls. Baseball veterans often say they knew they had to retire from the game when they couldn't hit the curve balls anymore. These baseballs don't overpower batters like blazing fast balls; they hamstring players who helplessly watch them dance across the plate—hooking, tailing, dropping, and twisting in such unbelievable ways that some batters are convinced the sharp drop of a curve ball is really an optical illusion or the result of the illegal use of sandpaper to scuff up the ball. Batters would like to believe that no human being could be talented enough to cause a leather-covered, five-ounce sphere to follow such an erratic course. It just ain't natural. The curve-ball controversy has been debated so intensely that in 1941, *Life* and *Look* magazines took stop-action photographs of curve balls to determine if the baseballs really did curve. *Life* concluded the "evidence fails to show the existence of a curve," while *Look* discovered just the opposite: The ball did curve. Even as recently as 1982, *Science* magazine commissioned scientists at General Motors and MIT to conduct a modern scientific investigation into the question, and once again stop-action photography was employed to show that a curve ball's curve is not an optical illusion but is based upon sound laws of physics.

Snapping Your Fingers

The secret of the curve ball is not its velocity (important to the fast ball) but its spin. Throwing a curve ball is a lot like twisting a doorknob or snapping your fingers. A sharp twist of the wrist puts a fast, forward spin on the ball. Like the dimples on a golf ball, the stitches on the baseball (all 216 of them) drag a thin layer of air around the ball.

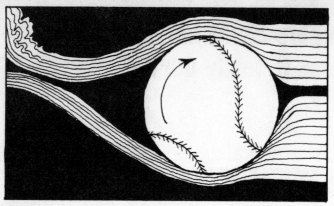

Illustration 34

Topspin of curve ball causes air pressure on bottom of ball to be less than air pressure on top, so ball sinks.

This topspin causes air to flow faster along the bottom of the ball than along the top. The faster-flowing bottom air is stretched thin, causing greater air pressure at the top of the ball and forcing the ball down (see Illustration 34). The curving force can move the ball down a foot or more in flight. If there were no gravity, the curve ball would form a circle two thousand feet in diameter. But, after all, the ball *is* moving on our planet and not in outer space. So gravity affects the path of the ball, too, pulling the ball toward the ground.

The force of gravity is a continuously accelerating force. It makes objects move faster and faster over time. So the effect of gravity is most pronounced in the second half of the ball's half-second flight to home plate. Combined with the curving force, gravity makes an overhand curve ball appear to drop suddenly, as if it had "rolled off a table" when in actuality the ball has followed a smooth arc during its entire flight (see Illustrations 35 and 36). So when batters say it appears that the ball makes a sharp break right in front of the plate, they are partially right. And when scientists say the ball follows a smooth, circular path all the while, they're correct, too. A pitch that takes less than half a second to reach the batter drops only half a foot due to gravity in the first half of

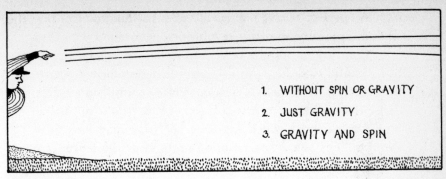

Illustration 35

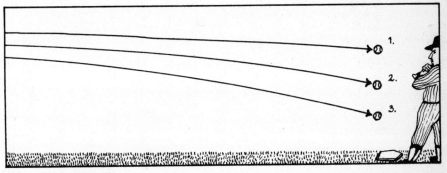

Illustration 36

The effects of gravity and the spinning ball combine to make a curve ball drop most near the plate.

flight, but in the second half it drops more than two feet. Is that a sharp break? If you're a batter, you think so. If you're a scientist, heck, no, but you've never had to face former curve-ball ace Sandy Koufax.

· ·

TRIVIA: Bottom (reverse) spin causes a fast ball to rise or "hop." Because the ball is spinning toward the pitcher, a fast ball's lower surface will be moving against the wind, creating greater pressure underneath the ball than above it, so the ball rises.

A "perfect" curve ball travels about a hundred feet per second, spinning at thirty revolutions per second.

Contrary to popular belief, a baseball travels farther in hot, humid weather. When the air feels "heavy" with moisture, it is actually "lighter." Hot, humid air is less dense than cold, dry air, so a baseball that might not make it out of the park when hit on a cool day may have just enough "legs" on a hot, humid one to clear the fence.

· ·

The Winning Edge

In skiing's early Golden Age
No arguments made skiers rage.
The ski was used for transportation
In every snow-bound northern nation.
—*Ski* magazine's *Encyclopedia of Skiing*

If baseball bats and tennis rackets have sweet spots, what secrets can we share with skiers? No need to leave skiers out in the cold; the secret to the ski is the edge.

Consider the first manuever every neophyte skier learns: the "snowplow." The snowplow involves joining the tips of the skis to form an inverted "v" and digging the inside edges into the snow. Anyone who has ever had a skiing lesson will remember the instructor taking pains to point out that the edges of the skis are the most important working surfaces and that by keeping the skis' edges securely dug into the snow, the movement and direction of the skier can be controlled no matter how difficult the mountain. A good snowplow can get a novice down the advanced skiers' Murderer Run as safely as if it were the beginners' Bunny Hill. All that must be remembered is to dig in the edges of the skis and to put pressure on them.

Why does this technique work? Here's the secret: The edges of skis are not straight. They may appear to be straight—sighted along an edge, they certainly look straight—but in reality each one has a slight curve; the side of the ski is slightly arced. This curve is actually part of a circle with a radius of 150 to 200 feet or more. It's this arc in the ski that makes it easy to turn. If you could somehow turn a ski on its edge, weigh it down slightly with a brick or two, and send it off on its own, the ski would not travel in a straight line but make a nice, long, arcing circle a thousand feet or more in circumference (see Illustration 37).

Illustration 37

The secret to the ski is its slightly curved edge, which would
cut a circle in the snow five hundred feet around.

Of course, the times when a thousand-foot turn are needed are
few and far between. Normally a much smaller arc, of just a few feet,
will do the job. That's where the ski instructor's advice pays off. When
one leans on the inside edge of the ski, the ski is bent, exaggerating
its natural curvature so the ski cuts a much shorter circle. Instead of
hurtling straight down the mountain at an uncontrolled speed, the
skier effects a nice, professional-looking turn as the edge bites into the
snow. Lean on the left ski and you turn right; lean on the right ski and
you turn left.

There's another pleasant consequence of this curvature. Lean hard
and long enough over the edge of the ski and what happens? The
turning circle will take you uphill! So to go *up* the mountain you have
to lean *down* the mountain—just the opposite of what your survival
instincts tell you to do but exactly what the skiing instructor has been
trying to drill into your head. No wonder novice skiers have such a
hard time sorting it out: They've been given seemingly contradictory
advice without ever being told why. By paying careful attention to the
edges of his or her skis, any skier, no matter how new to skiing, can
tackle the toughest hill. This, however, is a point lost on many novice
skiers who panic when looking down a steep mountain.

TRIVIA: For two thousand years, skis were nothing more than long slats of wood strapped to one's feet. After World War II, aircraft engineer Howard Head pieced together aircraft aluminum developed during the war and developed the first modern ski design: a sandwich—aluminum skins separated by a wooden core. It took Head almost six months to make the first ski. Today virtually all downhill skis are made of fiberglass. But the design has changed, functionally, very little over the past forty years despite the slickly colored materials and packaging.

The word *ski,* Norwegian for snowshoe, came from a Latin and Germanic word meaning splitting. It refers to the splitting of wood to form skis. Skiing may be more than five thousand years old. Skis dating back to 2500 B.C. have been found in Siberia. They look very much like those used today.

Flying Through the Water

One of my great loves in life is sailing. In fact, I owe my marriage to a class at a sailing school on City Island in New York where I met my future wife, Miriam. (City Island is New York's best-kept secret. Most provincial New Yorkers have never heard of it, let alone know of the excellent sailing facilities there.)

Illustration 38

Today not only are there sailboats of all shapes and kinds, some with one sail, some with ten or more, but also a new generation of "hull-less" sailboats (among the most popular). I'm speaking, of course, about Windsurfers. The most elementary Windsurfers are nothing more than surfboards with a sail and toeholds. Being lightweight and easy to transport, they're making sailing accessible to millions of people who could neither afford to buy a boat nor had the place to store one.

Despite their skimpy looks, Windsurfers, sailboards, and sailboats share the same sailing principles. Like most sailors, I had studied the theory of how to get a sailboat to sail—how to make sure the sails don't flap in the wind and how to turn and make sure I knew which way the wind was blowing (not an easy thing to do on a moving boat). But it wasn't until I had been sailing for a few years that I learned the secret of why a boat actually sails. By accident I found myself sharing a sloop with a sailboat designer who gave me a free science-of-sailing lesson that made me see the sea in a whole new light.

Faster Than the Wind

Like many novice seamen, I assumed that a sailboat gets pushed along by the wind, much like a bed sheet on a clothesline. When he

98

Illustration 39

asked me to explain how such propulsion could carry a boat up-wind, I casually pointed to the rudder and the keel and mouthed the wisdom that had been passed on to me over the years: The rudder pointed the vessel, and the keel kept it from tipping over. He accepted that answer for the time being, but then came the kicker. If a sailing vessel depended solely upon the wind to push it, then it would be impossible to sail *faster* than the wind. As iceboaters and yachtsmen know, it *is* possible to sail faster than the wind if given the right conditions. So there's more involved to sailing than just wind speed and direction.

Primitive sailboats had little maneuverability. They were little more than large rowboats with a single fixed sail that could be hoisted to catch a passing breeze and give the exhausted rowers a break.

Of course, this meant that your boat was held captive by wind direction; if you wanted to go north and the wind was blowing to the south, you either had to change your direction to match the wind's, or curse your luck and hope the wind direction would change to match yours. In about 3300 B.C. the Egyptians took a step toward solving the problem. They knew that if you stuck a large oar into the water at the rear of the boat, the direction of the boat could be changed. The Vikings took the solution one step further. They found that by

99

making their sail movable and by attaching a keel to its bottom, the boat could be steered even into the wind.

How much of the science of sailing could they have known? Did they understand the various forces at work pushing the boat in conflicting directions? Yet the Vikings sailed all over the world in ships designed without either the benefit of modern computers or the stellar mathematics of Newton's laws of motion. Their boat design relied on trial and error and an intuitive sense of nature. We're more fortunate because our science has discovered the secrets of the physics of motion and lets us enjoy them.

Sailboats That Fly

What the Vikings never knew but what we can show with lines and numbers is that a modern sailboat is really designed to "fly" through the water as much as an airplane flies through the air. Each part of a sailboat is aeronautically designed with this idea in mind. The sail, for example, is designed to act very much like an airplane wing set on edge. When a sailmaker makes a sail, he or she cuts the cloth and sews the pieces so that when the wind fills the sails they will have the shape of an airplane's wing—positioned vertically instead of horizontally. Some sails may be cut flatter and some fuller, but no matter what the sail, it will have a concave windward inside and a convex leeward (away from the wind) outside. Therefore, when the wind is blowing against the sail, the sail is not only being pushed in the conventional sense but also "lifted" like a wing (of course, in a sailboat, the lift acts horizontally to move the boat forward).

As it does over an airplane wing (see section "White-Knuckling It or Window Seat, Please," page 224), air splits as it flows over both sides of a sail. Because of the curvature of the sail, the wind flowing on the outer part of the sail has to go a longer distance than the wind traveling on the inside part. That means the wind must travel faster over the outer part to cover the same distance, causing the air to be stretched thin, which results in a drop of air pressure on the outside. This drop in pressure causes a lifting action in much the same way an airplane wing is lifted (sailors use the same term—lift—as pilots do).

The lift acts at every point on the sail surface, and the sail is cleverly cut in a shape that makes the suction greatest at the forward end of the sail (see Illustration 40).

The faster the air moves past the sail, the greater the lift. In actuality, the wind speed (expressed in knots) is converted into a force (measured in pounds). The faster the wind blows across the sail, the more pounds per square inch of lift are created. The force of the wind quadruples as its velocity doubles. If the water were to offer no resistance, the force on the sail would accelerate the boat to speeds much greater than that of the blowing wind. In some circumstances

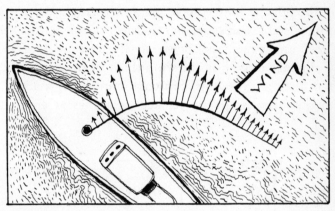

Illustration 40
The forward pull of a sail is caused by wind flowing over the sail, like air flowing over an airplane wing. The longer arrows represent the greatest suction.

the lifting action may account for up to 75 percent of the total force moving the boat ahead. This is the reason a boat can sail into the wind—it is being lifted rather than blown. Of course, no boat can point directly into the teeth of a wind and expect to make headway. About as close as the average boat can get is to sail within 45 degrees of the direction from which the wind is coming, on either side of the wind's eye.

Only the most well-designed and well-handled racing boats can sail any closer to the wind, perhaps up to an angle of 35 degrees. So to head upwind, boats must tack or zigzag their way from point to point, always changing direction to keep the wind blowing over one side of the boat or the other.

The Flying Keel

When blowing against the boat and sails, the wind also pushes the boat sideways. This sideways thrust is very strong and not desirable. It will throw the sailor off course and if not counteracted in strong winds, the thrust will make the boat very difficult to steer and might even tip it over. That's where the keel comes into play. The keel takes advantage of this sideways slippage ("making leeway"), and by "flying" through the water much like an airplane wing it helps make the boat more stable and steerable. Consider the view from the bottom of the keel underwater as shown in Illustration 41. Due to the sideways slippage caused by the wind the boat never moves exactly in its desired direction (heading). Instead it moves at a slight angle to that direction (yaw angle) so its keel is always at an angle to the heading.

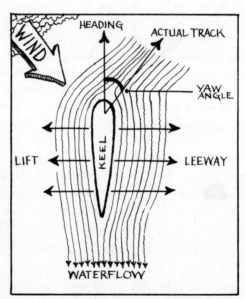

Illustration 41
Keel Heading versus Actual Heading shows that a sailboat is always slipping just a bit sideways as it moves forward.

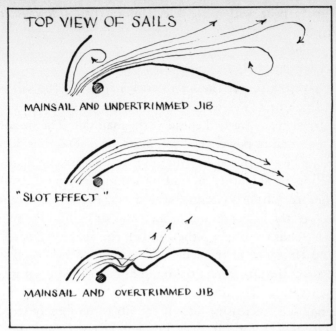

TOP VIEW OF SAILS

MAINSAIL AND UNDERTRIMMED JIB

"SLOT EFFECT"

MAINSAIL AND OVERTRIMMED JIB

Illustration 42
Slot effect. Sketch of three boats (top, middle, and bottom)
looking down from above. Boat (top) has jib and mainsail
showing jib eased too much. Boat (middle) has jib correctly
trimmed, captioned "slot effect." Boat (bottom) shows jib
tightly trimmed, "jib backwinds mainsail." Shaded arrows
show wind passing through sails.

Because of yaw angle, water rushing past the keel does not flow
evenly around both of its sides. Water on the windward side has a
farther distance to travel than water on the other side, producing an
area of reduced pressure resulting in that now-familiar lifting or
suction force. The more underwater lift, the greater the force resisting
sideways slippage.

In actuality, the force moving the boat is a combination of
components that push the boat both forward and sideways. And the
entire boat, from the sails at the top to the keel at the bottom,
functions as a wing. In fact, if it were not for modern hull design the
boat's primary motion would be sideways. The flow of water over the
entire underwater surface provides a certain amount of life: Even with

no sails up, a boat will move ahead more than sideways in a side (beam) wind.

Increasing Lift

Is it possible to get the air to suck harder on the sail? Yes, by putting a second sail in front of the first. On many boats you'll see a smaller sail, the jib, attached ahead of the mainsail. The size of the sail makes it too small to power the boat on its own, though it does supply some drive. The real purpose of the jib is to increase the mainsail's lift by channeling air smoothly around its winglike surface. As shown in Illustration 42 (slightly exaggerated for clarity), when the jib is adjusted correctly, it bends and funnels the air behind the main. This "slot effect" tends to speed up the wind on the mainsail, increasing the suction and efficiency of the sail.

The faster the air travels, the more difficult it is to get it to follow the sail's curvature. The jib also acts to shape the wind and forces it to hug the mainsail, increasing lift. If the jib is too loosely trimmed, as in Illustration 42, top, it does not effectively create the "notch" between itself and the mainsail. The boat loses speed. If the jib is too tightly trimmed, as in Illustration 42, bottom, it "backwinds" the main, creating turbulent air that disrupts the slot effect. Jibs come in many sizes. The larger the jib, the better the air is shaped to follow the curve of the mainsail and the greater the drive. Though the boat can sail with just the main or the jib, the combination of *both* adds up to greater efficiency than the sum of each acting alone.

• •

TRIVIA: Sailing is like squeezing a watermelon pit. When the pit is squeezed between the thumb and forefinger, it shoots forward. Similarly, when a sailboat is caught between the squeeze of the wind on the sails and the water on the keel, it is propelled frontward.

• •

The Hand Is Mightier Than the Board

People like to hit things. Baseballs, tennis and golf balls, and each other. Most of the time objects—bats, rackets, boxing gloves—are used as tools of attack. But in some sports people use nothing more than the naked parts of their bodies. In fact, some people are so talented at this feat that others will pay them to do it in public. We call this karate, the art of breaking hard objects with parts of your body.

At first glance, karate appears to be a highly masochistic sport. Why would anyone deliberately want to court injury by striking a solid object?

Kara-te means both "empty hand" and "China hand." The roots of karate can be traced back to the seventeenth century. Its earliest form, called *te,* or "hand," was used by the Okinawans as a means of fighting an occupying force of Japanese. After having been disarmed by the Japanese, the Okinawans turned their hands and feet into weapons strong enough to smash the samurais' bamboo armor. Although Okinawan resistance failed to drive out the Japanese, the art was passed down through the generations.

Karate looks like magic only if you don't know the secret: physics. The secret of karate involves three factors: speed, mass, and the rigidity of the weapon—your body.

Why speed? Consider this case: I am holding a board you are to strike with your hand. To break the wood, the duration of impact has to be very short. Hitting the board very slowly would just push it away. But a sharp, short thrust would give the board no time to recoil, and you'd chop it in two.

What about mass? For a few boards, a hand will do fine. But the more boards to break, the more massive the part of your body you should use. To karate-chop four or more boards, a foot works much better. The speed of the impact remains the same, but a foot is much more massive: More force is delivered. A foot imparts greater crunch per punch.

The third factor is rigidity. Your hand has to be as rigid as possible at the moment of impact. A tight hand will absorb less of the energy, so more of the energy will go into breaking the wood. Notice I said "hand." We novices do not break boards with our fingers. Even though the "chop" makes it look as if the fingers are delivering that withering blow, they have no involve-

Illustration 43

ment. It's the fleshy part of the hand, between the little finger and the wrist, that does the work. As a novice, you literally make a fist and bang your hand down through the board as if your fist were a gavel. Your fingers stay out of the picture.

The source of the power behind the chopping motion is familiar to anyone who watches baseball. Just as a pitcher uses his whole body to gather up speed for a fast ball, so a karate expert uses his entire frame to deliver the speed of the blow through a small part of the hand (or foot, or head, etc.). Done correctly, no part of the body is injured, no bones broken. The hand is very elastic, so it bends. The board isn't, so it breaks.

Bones are about sixty times more elastic than pine boards. It takes about two hundred pounds of force to break a pine board. A hand is capable of withstanding about a ton of force (two thousand pounds). So theoretically you should be able to break at least four or five boards with your hand before your hand is hurt.

· ·

TRIVIA: A karate blow must be extremely short and swift to be effective. To break a brick, a hand should be in contact with the brick for only five to ten one-thousandths of a second. Any longer and the hand will dampen the vibration needed to crack the brick. This means that after impact, the hand must recoil at almost two hundred miles per hour.

· ·

IV. Kitchen Magic

I feel a recipe is only a theme,
which an intelligent cook can play
each time with a variation.
—Madame Benoit

When I was a boy, my mother was concerned about my love of the kitchen. "Why do you spend so much time in the kitchen?" she'd say. "Kitchens are for girls. I'm worried about you."

It's not that my mom didn't understand that kitchens are natural places to hang out. We all spent time there. But I had taken a peculiar interest in the food. I reveled in watching the cooking process, in standing over my mother's shoulder (via a chair) and staring into the pots. To me the kitchen was more than just a social hall. I was fascinated by the magic of the kitchen and wanted to understand the changes food undergoes when it is cooked.

Why does raw food turn color? Why does water bubble when it boils? Why does egg salad spoil on a picnic?

Most people don't stop to think that when they poach an egg or broil a steak, they're actually conducting a culinary experiment being repeated the world over. The kitchen is a domestic laboratory, and in comparing one soufflé with another, it's obvious some of us are better chemists than others. If my skill as a cook has improved, a lot of credit goes to understanding the chemical and physical changes occurring in food when the ingredients are mixed, cooked, and served.

. . . Cauldron Bubble

The worst thing you can say about a cook is that he or she can "burn water." That's some insult, considering the impossibility of the feat: Water can only boil; it never burns. Boiling is the easiest, simplest, and most basic form of cooking, and for my money the most mysterious because anyone who has ever watched a kettle knows that boiling is not an easy event to understand. That's because lots of fascinating changes are occurring inside that spaghetti pot or tea kettle.

109

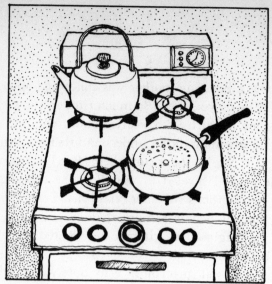

Illustration 44

Place a pan of water on the heated burner and what happens? Nothing immediately, but soon the pan begins to "hiss" or "sing" for no visible reason. Little bubbles begin to form and rise to the surface. Soon the water begins to bubble noisily, and if you are an inexperienced cook (or a lousy physicist) you might think the water was ready for tea. But if you look closely you'll see that the bubbles rising from the bottom do not reach the surface. They appear to "dance" on the bottom and disappear on the way up. Eventually the bubbles rise higher and higher until, like a ladder, they reach the surface. At the same time the loud bubbling has leveled out to the smooth, soothing sound of a full boil. The water is now ready for the tea bag or pasta.

What is going on here? Why does the pan "hiss" when I put it on the fire? What are those little bubbles at the bottom? How do I know when the water has boiled "long enough"?

Boiling is not the same as evaporation. Evaporation occurs only at the surface of a liquid when water molecules escape into the air as a gas called water vapor. Water molecules escape because they are always moving. Some move faster than others; their speed depends on their temperature. Warmer molecules move faster than colder ones. Some of the molecules at the surface of the water are moving fast enough to puncture the surface tension of the water and escape into the air. That's the essence of evaporation.

Boiling is really a special case of evaporation. Boiling is the state in which water vapor escapes from *all* parts of the water, the bottom and the middle as well as the surface. Normally water starts boiling at the bottom, where it's closest to the heat. The heat agitates water molecules, giving them enough energy to leave the liquid state and turn into steam. But in the early stages of heating, as the steam bubbles rise to the top, they meet colder layers of water. The bubbles are cooled, the pressure inside the bubbles collapses, and the bubbles die. The steam is converted back into liquid water. The bursting bubbles don't die quietly. Like tiny balloons, they bang when they burst; and when heard together, these collective pops make the chorus of noise heard before the water has come to a full boil. Only when the upper level of water has been heated to the boiling point—212 degrees Fahrenheit (100 degrees Celsius)—do the bubbles survive and form a chain from the bottom of the pan to the top. The water is now at a uniform temperature and is properly boiling and ready for your tea bag or pasta. At the boiling point the pressure of the escaping steam is equal to the normal atmospheric pressure of the room. That's why the water now escapes so easily.

Even the full boil state confuses people. They think that by turning up the flame, the water can be made hotter to cook food faster. The logic is sound but the physics isn't. Once water starts boiling, it has reached its final temperature. You can turn the range up all you want, but you won't make the water hotter. Why not? When water turns to vapor, it cools itself off. Remember that when water vaporizes, the most energetic molecules escape first. That leaves the sluggish, cooler ones behind, lowering the temperature of the water. (This is the same reason you feel cold when you step out of the shower. The evaporating energetic water molecules leave the cooler ones behind on your skin.) The flame applied to a boiling pot of water does nothing more than counteract the cooling process by replacing the lost heat. There is no more heat left to raise the temperature of the water. Yes, by turning up the flame the bubbles will become more violent, but that just means the water is boiling away faster—not becoming hotter.

There *is* one practical way to increase the water temperature: make it more difficult for the water to boil. How? Put a lid on the pot.

111

The old pressure cooker does just this. With a tightly capped pot, steam accumulates under the lid and builds up pressure on top of the liquid. More pressure means those water particles need a lot more energy to escape from the liquid. (Just try to jump out of a swimming pool with someone on *your* back.) More energy is needed by water molecules to make the jump from the water into the pressure chamber, and that means that more heat must be added. That explains why pressure cookers cook food so much more quickly: In a pressure cooker the temperature of the water can be increased to 250 degrees Fahrenheit. Broccoli, normally boiled for about eight to fifteen minutes, takes only ½ to 1½ minutes in a pressure cooker.

Of course, the opposite is also true. High in the mountains, where the air is thin, water boils at a lower temperature because atmospheric pressure is lower. In Denver, 5,280 feet above sea level, the boiling point is as low as 203 degrees Fahrenheit. Move to Mexico City's lofty altitude of 7,347 feet and you've cooled the boiling point to a chilly 198 degrees Fahrenheit. Chemical reactions that cause food to cook are sensitive to heat. Cooking time is decreased by a factor of ten. Of course, out in space, where there is no air pressure to hold in the vapor, water boils instantly. The water molecules don't need much help (in the form of heat energy) to escape from the liquid. So even cold water boils. That's why astronauts wear space suits. You can imagine what would happen to their blood if they didn't.

A question for the day: Why can you safely stick your hand in a hot oven (400 degrees Fahrenheit) for a few seconds to remove a roast, while dipping your fingers for just a fraction of a second in boiling water (at a cool 212 degrees Fahrenheit) will give you a bad burn? (Even science teachers have trouble with this one.)

Water is about a thousand times denser than air. Therefore, many more water molecules strike your skin in boiling water than do air molecules in the oven, so your skin gets a bad burn. It's akin to being hit with a single Ping-Pong ball or a barrage of them. One ball doesn't hurt very much, but a steady stream can be quite painful.

Illustration 45

TRIVIA: Mountain climbers can boil their food for hours without cooking it. At the extremely high altitude of Mount Everest (29,028 feet), water boils at about 167 degrees Fahrenheit—hardly hot enough to cook food or make a decent cup of coffee.

Eggs-quisitely Done

Illustration 46

I would guess that eggs are among the most commonly consumed foods, eaten either by themselves or in other dishes. Soft-boiled eggs are usually too runny, hard-boiled ones crack and turn green, fried eggs are burnt around the edges, while the yolks stubbornly remain uncooked. These failures are frustrating because most people are very fussy about their eggs. Grown cooks have been known to cry over a wounded sunny side up, or an egg shell that refuses to be peeled. There is something about cooking the perfect egg dish that brings out the compulsiveness in even the most carefree chef. Eggs have to be done just right.

With a little understanding of chemistry and physics, you need suffer no longer. The perfect egg is as close as your condiments. The first course in eggs-quisitely done eggs involves understanding the egg itself.

Nature packages eggs conveniently. Neatly wrapped in a hard, protective shell, they last a long time. An egg has an efficient structure designed to protect the incubating embryo from harm and to give that

114

chick nourishing prenatal care. Standard equipment includes an air sac "shock absorber" at one end to fend off vibrations; a soft clear "white" made of protein; and a fatty "yellow" full of cholesterol, which may not be good for us but is an essential ingredient for the development of cell tissue in the growing chick. And herein lies the first roadblock to the perfectly cooked egg: the protective "air sac" at one end. This air sac is responsible for one of the greatest hard-boiled-egg tragedies: ugly white streamers.

Prick the End of the Egg Before Boiling

When cooking a hard-boiled egg, the air sac in the egg expands as it's heated. Notice the little bubbles coming from the large end. That's the egg trying to let the expanding air seep out through pores in the shell. But if the pressure builds up too quickly, the shell cracks, releasing egg white into the water. The remedy: Prick the end before immersing it in water to let the air out and help prevent cracking. As an added bonus, pricking the end also makes a hard-boiled egg more appealing: Releasing the pressure gets rid of the "flat" part of a hard-boiled egg. The size of the air sac is a clue to the freshness of the egg. As an egg gets older, the air pocket gets bigger. For those who don't trust their abilities to prick an egg without cracking it, start the egg in cold water. The added cooking time gives the air more time to seep out.

Too Late: It's Cracked

There comes a time when even the most experienced eggs-pert cracks a shell. What follows can only be described as an ugly mess as egg white seeps out of the crack and fills the pan with stringy white streamers. This is no cause for alarm. Put a little vinegar or salt in the water. The salt or vinegar (acid) releases electrical particles into the water that coagulate the white and make it cook much faster. So the crack is sealed up quickly by its own escaping egg white. *Voilà!* No ugly egg streamers! For the same reason, great chefs make poached eggs in vinegared water: The acid helps cook the white quickly so the poached eggs stay in appetizing, compact little balls that sit nicely on a piece of toast. (I'm getting hungry.) Ever wonder why poached eggs are served with a slice of orange or pineapple? Now you know: to mask the slightly acidic taste of the poaching water.

The Forbidding "Green Yolk"

It's extremely embarrassing for a cook who has spent the entire afternoon getting ready for a dinner party to find that the yolks of the hard-boiled eggs have all turned green. Of course, there is no danger in eating these eggs; they merely look unappetizing. But with a little knowledge of kitchen chemistry this problem need never plague you. Simply avoid boiling the eggs too long, and when they are done, run them under cold water. That should keep them nice and yellow. Why?

As an egg cooks, some of the proteins in the white breaks down and releases hydrogen and sulfur, which combine to form hydrogen sulphide gas, better known as the substance that causes the smell of rotten eggs. This gas goes to the coolest part of the egg, the center, where the yolk is. The yolk is full of iron, which has a great affinity for the gas, or rather, for the sulfur in it. The iron and sulfur unite and give birth to iron sulfide—the ugly green stuff on the surface of the yolk.

To prevent this unwanted offspring, just cool the eggs quickly under cold water.

Keeping Fried Eggs from Burning Around the Edges

Repeat after me: "Turn the heat down." Overcooking destroys wonderful fried eggs. Why?

Raw egg protein—egg white (88 percent water, 11 percent protein)—normally exists in a ball, closed off by weak chemical bonds from the other protein molecules balled up nearby. Once its temperature is raised, water molecules energized by the heat go into a frenzy, bouncing off one another, knocking into everything in sight, including the protein. Unable to remain tightly ball-shaped amid the chaos, the protein molecules unravel into long streamers. Whereas the coiled protein molecules were solitary and inhibited, the uncoiled streamers immediately find one another and bond, forming strong links. More and more join together, forming a network of streamers that feel smooth and springy to the touch. This is the cooked egg white. (Salt and acid have the same uncoiling effect. Remember why they were added to the cooking water?)

116

When things get too hot, the situation gets out of hand. No more smooth linkups occur. Just molecules of rubbery, ugly stuff. Water in the egg is immediately vaporized, drying out the white. Then, as the eggs continue to be heated, matters get worse: The edges turn dark and crusty. Sugar and protein in the egg white combine. Once this "browning reaction" takes over it's a sure sign this egg won't be served to guests (maybe to your mother?). So keep the heat down. The browning reaction takes place only at high temperature.

"But the egg is still runny. The white's not done!"

Give the egg a steam bath. Put a tablespoonful of water in the pan and slap on the lid. There's nothing like a little steam to bring heat from the bottom of the pan to the top of the egg.

When asked how you learned to make such beautiful eggs, just say you took Chemistry 101.

· ·

TRIVIA: Eggs are roundish for a good reason. The oval shape makes the shell very strong. Squeeze the ends of an egg between the palms of both hands. It's very difficult to break the egg this way. The rounded ends are very strong, like the dome of a building or the archway of a wall. That's why eggs are stacked on end. **WARNING:** Don't squeeze the middle of the egg unless you want an instant omelet.

If your doctor puts you on a one-egg-a-week diet, then an ostrich egg is for you. One 3¾-pound beauty is enough for about two dozen breakfasts. Allow forty minutes to boil.

Mayonnaise does not spoil sitting out in the heat. If egg salad or chicken or tuna salad spoils on a picnic, it is not the mayonnaise that is at fault. Mayonnaise is made with eggs and vinegar. The vinegar makes mayonnaise so acidic that food poisoning germs cannot grow. A jar of mayonnaise can remain unrefrigerated for days without spoiling. But when food is added to the mayonnaise, lowering its acidity, germs can feast on the food. Leaving the food in the hot sunlight creates the ideal warmth microbes need to multiply.

· ·

Don't Burn Your Tongue!

I was never quite sure if blowing on food to cool it off was an inborn trait, inherited through the genes, or something that had to be learned. It is done so commonly all over the world that one is tempted to imagine how the habit originated.

Somewhere, millions of years ago, just after learning to use fire, the primitive human was faced with a difficult technological dilemma: how to cool his piping-hot food so he could eat it. Surely he must have burned his tongue enough times to wonder if this cooking business was worth the effort. Cold food, after all, didn't bite back. And before he could spread this new technology to his friends, he would have to find a way of eating the hot stuff or he wouldn't have many friends left. Like any technologist working on the cutting edge, this culinary Cro-Magnon must have experimented with many different cooling methods: dancing around the fire, chanting over the bowl, even dousing it with water. Perhaps water was the initial solution, but an unsatisfying one—a soggy mess is not something you're proud to serve to company. There must be a neater, quicker solution. Sooner or later he must have discovered that by pursing his lips and exhaling across a steaming bowl of mastodon stew, the food magically cooled. Why? He hadn't a clue, but it worked.

This technique was passed down from generation to generation. No one has found a genetic marker to indicate if it was incorporated into our chromosomes as a survival technique (but has anyone looked?), and after watching my two-year-old son, Sam, struggle with the problem, I'm convinced it is an acquired habit. But even Sam doesn't know why he does it, only that "blowing on it" works. This explanation is sure to be the talk of any dinner party when the soup is served.

Most people would say that blowing cools the soup because your

Illustration 47
Primitive humans contemplate how to cool hot food. How did they discover that blowing on it cools it down?

breath is colder than the liquid. That's true but certainly not cold enough to make much of a difference.

The real reason that blowing cools hot soup and other foods has to do with evaporation. In discussing the process of boiling (see ". . . Cauldron Bubble") it became apparent that evaporation is a cooling process. For molecules of water to evaporate, they must literally be made to jump out of the water. So only the most energetic ones get out, leaving the slower, cooler ones behind, the result of which is to lower the temperature of the whole liquid. The more quickly the liquid evaporates, the faster it cools.

That brings us back to our soup. If the food is hot, it aids in the evaporation process. So the soup should be cooling quickly. But it doesn't. Why not? Consider the microclimate above the bowl of soup. With all that vapor rising out of the bowl, the area above the soup forms into a hot, steamy cloud. The air is almost saturated with evaporating water. The evaporation process slows down to a crawl and so does the cooling. The vapor just sits over the bowl, hanging still and heavy, like the humidity on a hot, windless summer day. The soup will take a long time to cool off.

119

What we need is a refreshing breeze. By blowing on the soup we disperse the vapor and bring in fresh, unsaturated air ready to absorb the vapor from the evaporating liquid. The evaporation continues rapidly, and our soup cools quickly. (Of course, if the bowl is too big or the dinner party too elegant, one can blow on the soup in the spoon instead. This is the more commonly accepted practice.)

Illustration 48

Unless, of course, we're eating a fatty soup. Ever notice how fatty chicken soup or meat broth takes forever to cool off? There's a good reason for this. The soup has a lid on it. The layer of fat floating on top of the soup, though very thin, acts to prevent the evaporation of the liquid. Since the fat itself will not evaporate, the soup is virtually cut off from the air and evaporation is halted. The only solution is to remove the fat or break up the impenetrable skin. Some people skim the oil off the surface; others just stir the soup over and over again, not giving the fat globules enough time to collect and link up on the surface. (When eating fatty soup, blowing on the soup, spoonful by spoonful, is almost mandatory.)

Stirring is always a good cooling method, whether it be soup or coffee. By stirring we bring up hot liquid from lower down to replace the cooler liquid at the surface. This speeds evaporation.

Of course, the opposite is also true. To keep something warm, always put a lid on it. Your dinner guests might say that a lid keeps the heat in. That's only partially correct. The lid locks in the water vapor rising from your soup or boiled potatoes and prevents it from escaping from the serving dish. So inside the covered dish the air is saturated with water and the evaporation process has slowed down drastically. The steamed vegetables stay warm. Yes, a certain amount of heat escapes through the walls of the dish, but the majority of heat lost through the cooling of hot, moist food is lost by evaporation. So instead of merely taking food out of the oven to cool, it's better to take the lid off at the same time and let the vapor escape.

120

Fire: It's a Gas

Fire is hardly new. But ask anyone standing over a hot gas burner or huddled close to a fireplace just what fire is, and you're sure to receive a blank stare. It's not the kind of question answered in Trivial Pursuit. For many years, in an attempt to understand fire, alchemists debated the issue of whether or not fire was alive. This idea is not as silly as it sounds. Consider all the properties fire has: It eats, breathes, gives birth, and moves from place to place. These are all qualities of living things. So why isn't fire alive? Because it lacks substance—protoplasm is the scientific word. If it weren't for this small detail, you could probably build a good case for fire being alive.

So what *do* we mean when we say something is *burning* or has caught fire?

By burning—combustion—we mean a rapid combination of a substance with oxygen via a chemical reaction. The most important fact to remember about burning is that nothing will burn in a solid form. It must first be converted into a gas or gases. When a match is held to a candle, before the wax can ignite, it must be vaporized. The same is true for the logs in your fireplace. They must be turned into a gas by burning twigs before they will catch fire.

That's where heat comes in. Consider our candle. In its solid form, the wax cannot rapidly combine with oxygen. But the heat of a match drives off combustible gases into the air, giving them a better chance to react with the oxygen, gas to gas.

So a flame is merely burning particles of gas.

Some things, like gasoline, don't have to be heated very much; they vaporize easily without assistance. Just open a can of gasoline and watch the fumes (vapors) come out. A piece of coal, on the other hand, has to be heated quite a lot to be turned into a gas that will burn.

Most substances that burn contain hydrocarbons—carbon and hydrogen—which combine with oxygen to produce carbon dioxide

Illustration 49

(CO_2) and water. These end products have much less energy than the burning hydrocarbons so the excess energy is released as heat. The resulting heat rips apart more fuel to produce more burnable vapors in a process that goes on and on until the fuel runs out.

Air Pollution and Rusting Apples

What is smoke? Smoke and other pollutants are made of incompletely burned particles. A clean-burning fire won't produce any smoke. Your gas range is a good example of that. The gas burns completely to produce just CO_2 and water. There is no smoke and little if any of the incompletely burned carbon that produces lethal carbon monoxide gas. Gas ranges don't have to be vented to the outside like wood-burning stoves because gas burns more cleanly.

The very hot and rapid burning of a lighted candle produces a visual display we call a flame. The rapid disintegration of the gaseous wax and recombination into water and carbon dioxide involve a lot of energy. Our bodies do this with the aid of enzymes, at a much lower temperature. But results are the same: carbon dioxide and water vapor

122

you can "see," exhaled, on a cold day. Observing a half eaten apple is a good way to see slow burning. The "brown" flesh is the product of the apple combining with oxygen in the air. Rusting involves the same effect with iron—it would not be incorrect to say that the apple was rusting.

Flashover

If a fire should start in your house, the gases produced by heated furniture are most dangerous. In addition to being poisonous, these vapors collect in a cloud at the ceiling. The gas cloud can get very hot—up to 1,200 degrees Fahrenheit—and when it reaches that point a very unusual and lethal event occurs: flashover. The heat from the gas cloud causes more vapors to come out of the floor and the furniture. The vapors violently ignite, bursting the whole room into flames. That's why it's so important to get out of a burning building as fast as you can. Don't be a hero and go back for a pet.

· ·

TRIVIA: Gasoline will vaporize even at −45 degrees Fahrenheit. Since you cannot see the vapors, gasoline and other liquid fuels will ignite quite a distance from their source. These vapors hug the ground, where a lit cigarette or errant spark will ignite them and send a trail of flame rushing back to the open liquid many yards away.

All chemical reactions in a flame take place in a thin, bright cone called the flame front. The reactions move across the flame front at speeds of more than three thousand feet per second. The tearing and welding of elements inside a flame are so fascinating that in 1860 the great English scientist Michael Faraday devoted an entire Christmas lecture for children on the topic "The Chemical History of a Candle." It was standing-room only. Each night for six days a rapt audience of over five hundred, adults included, listened intensely as Faraday depicted the chemical beauty of a simple flame.

· ·

Nuke It in the Microwave

My wife and I had just finished picking the appliances for our new kitchen. Our old kitchen ran quite well with the simple amenities—stove, oven, and refrigerator—and I was struck by how little modern kitchens had changed over the thirty years since they were first installed. Except for one important factor: the oven. Microwave ovens have revamped not only the way we cook but also how we run, work, and plan our days. Working men and working women have become liberated. No longer must an exhausting day on the job lead to an entire evening in the kitchen. A frozen dinner "nuked" in the microwave turns an hour into minutes. As for the unmarried, they bless the microwave every day. No need to heat up a big conventional oven to cook a lonely meal for one. A microwave fit snugly under the cupboard does the trick.

In hot climates microwaves do not heat up the kitchen the way a conventional stove does. A microwave oven is perfect for those in wheelchairs who can install the oven at a convenient height.

The ovens are not without faults, however. They won't brown foods easily, lots of practice is required to use one well, and a microwave bulging with a five-course meal will take just as long to cook as it would in a conventional oven. But once they try a microwave oven most people are hooked, and microwave users will be happy to learn that some ovens now being produced are hybrids—microwave/convection combinations. These hybrids overcome the shortcomings listed above.

No Nukes

No matter what people say, "nuking" a baked potato does not mean heating it by nuclear radiation. There is no little nuclear reactor in your oven. Microwaves are not radioactive (technically they're called "nonionizing" radiation). Microwaves belong to the same family of radiation as light, radio, television, and radar. The name "microwave" stems from their size—much shorter in length than the electromagnetic waves used to broadcast radio and television signals. In a conventional oven, food is heated by a combination of methods: radiation, conduction, and convection. Heat rays from the glowing element sear the food, hot air in the oven helps browning, and heat from the dish make its own contribution.

Microwaves concentrate all their energy at a single frequency and to a great extent heat food, water, and baking dish solely by radiation. The most effective target of microwaves is water, the most abundant and important ingredient in food. Water molecules are dipoles. This means that they act like little magnets having a plus charge on one end and a minus on the other. No one knows the exact nature of the heating effect produced when microwaves strike water molecules. But it's suspected that when electrical energy in the form of microwaves strikes the water molecules, it causes a rapid twisting and turning of the molecules, resulting in the production of heat. This is the major action responsible for microwave cooking, and it explains why microwave ovens heat food rapidly: The power of the microwaves is converted directly into heat within the food. This also explains why the oven stays cool—the microwaves give up their energy not to the walls of the oven but only to the food. Microwave cooking is very efficient; up to 50 percent or more of the microwave energy is turned into useful cooking heat.

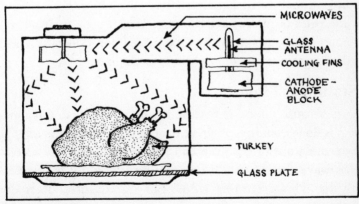

Illustration 50

The Great Microwave Myth

No one knows how it all started. But about twenty-five to thirty years ago, when microwave cooking was in its infancy, the word got out that microwaves "cooked the food from the inside out." Nothing could be further from the truth. Microwaves do not pierce very deeply into food. In a roast beef, for example, microwaves penetrate only about an inch or so before most of them are used up. The very center of the meat is never touched by microwaves, and it cooks in very much the same way as it does in a conventional oven: by the slow transfer of heat from the outside inward.

In a conventional oven this process presents no problem. The slowness of the heating method allows plenty of time for heat to work its way through the meat, so ultimately, there will not be a great difference in temperature between the outside and the center because the meat has had plenty of time to evenly transfer the heat. But a large piece of meat can pose a major problem in the microwave. Due to the rapid cooking ability of microwaves, the outside of the meat will cook quickly while the inside stays almost raw.

The simplest way to overcome the problem is to turn down the power and slow the process to allow heat from the outside to seep into the center. But doing this defeats the whole reason for using a microwave oven: speed. Lowering the cooking power lengthens cooking time to the point where one may as well cook the roast in a conventional oven. As a rule of thumb, any food that takes forty minutes to cook correctly in a microwave might just as well be cooked in a conventional oven. An alternative suggestion is to follow one of the prime rules of microwave cooking: Under cook—always take food out before it's done—so it can be tested for doneness and can be allowed to continue to cook while it is cooling.

Standing Time

Just because a food is removed from the oven doesn't mean that the food is done cooking. A great deal of cooking occurs after the food has been taken out of the oven and left to cool. This is especially true for microwave ovens. While the outside is cooling off, the center is heating up. This "standing time" can significantly affect doneness, especially when cooking roasts. Helen J. Van Zante of the University of Wisconsin studied what happened to a four-pound meat loaf when microwaved and left to cool. She found that the internal temperature of the meat loaf when removed from the oven (160 degrees Fahrenheit) continued to rise until it reached a peak of 180 degrees Fahrenheit ten minutes later. After forty minutes of standing time, the inside of the ground meat was hotter than at the end of forty minutes of cooking time.

Food Shape Makes a Difference

Because of the behavior of microwaves, the shape of the food plays a key role in how fast and evenly it cooks. Since microwaves have a tendency to bounce off bony objects, cooks will find surprising things happening in microwave ovens. Consider cooking a whole turkey versus a rolled turkey breast. Which will cook faster? Surprisingly, the whole turkey. Whole turkeys are hollow, and microwaves like to bounce around inside the cavity, ricocheting off the bones. Pointed objects are like microwave lightning rods: Microwaves are attracted to them. So thin, pointy objects, like turkey or chicken wingtips and drumsticks, need to be shielded with small pieces of foil, or they will get burned.

Because shapes are important, deciding which shape to use will affect cooking time. A rolled rib roast will cook differently from a standing rib roast. Even the length of rolled rib roast will make a difference. A short, fatter one will take longer to cook than a thin, longer one and the final product will be more difficult to control.

A Dangerous Cup of Tea

Many people heat a cup of water in their microwaves to prepare a cup of tea. They set the timer, and when it rings they dutifully drop in a tea bag with alarming results: the water appears almost to explode, coming to a rapid, violent boil and scalding anyone gripping the cup. I've even had this happen to me when heating a glass mug of water. This "instant boil" occurs because microwaves superheat the water—that is, they heat the water higher than the boiling point without any visible signs of boiling. How does this happen? Microwaves heat the center of the water so quickly that the water near the sides and top of the cup remains cool. But when a tea bag is dropped into the water, bubbles of water vapor form rapidly along the jagged, porous netting of the bag (see explanation of boiling in ". . . Cauldron Bubble"). The water will explode in a brief, rapid boil until the temperature drops.

There have been reports that a spoon or granulated sugar placed in a superheated cup of water can also start the rapid boiling.

Uniform Defrosting

Microwaves penetrate deeper into frozen foods than thawed foods. The frozen water molecules, locked in their crystalline form, cannot be twisted by the microwaves, so they do not immediately heat up. But since frozen food is never completely frozen, there are always tiny sites of liquid water that absorb the microwaves, heat up, and begin the thawing process. Left to themselves, these pockets of water would quickly cook the food around them while leaving the rest of the food frozen.

Also, when a roast is taken out of the freezer, the outside surface immediately begins to melt. In the oven, a layer of liquid water on the meat's surface will sponge up microwaves while the inside of the meat will let them pass through.

That's why thawing is best done at low power or on a special thawing cycle that periodically turns the power off. This way the heat is uniformly conducted to the frozen sections so the food can be evenly thawed.

Container Shape Makes a Difference

Laboratory tests show that round pans are better than square ones. Square corners provide a convenient place for microwaves to concentrate. The corners get rays from three directions; as a result, they heat up quickly. Wise cooks will shield these corners with aluminum foil. The depth of the dish also matters. In many cases a shallow dish will heat food faster than a deep one. But the large surface area of a shallow dish allows lots of steam to escape, and since microwaves love water, heating power that might otherwise be going into the food is being wasted on the water vapor in the oven. Veterans of microwave cooking know always to cover their dishes to avoid loss of water, which lengthens cooking time.

Test Your Metal

Metal objects of all kinds—from aluminum foil to pots and pans—are usually forbidden in a microwave oven. One need not observe this rule religiously, however. It is true that metal reflects microwaves like a mirror reflects light, and with no place to go the reflected waves can overload the oven and cause damage to the microwave generator called the magnetron. It's also true that microwaves have a hard time penetrating metal containers to reach the food placed inside them. But metal foil can be used sparingly in special situations as a shielding to reflect microwaves: on wingtips and on drumstick ends of poultry, over the rib eye of meat, and on the corners of dishes. Twist ties should be removed because their sharp metal points may cause sparks.

Not all nonmetallic materials make good containers. The ideal container will allow microwaves to pass through but will not absorb or reflect them. There's a simple test that will determine if a nonmetallic utensil is microwave-safe. Place the empty container alongside a glass of water in the oven (the water acts to absorb stray microwaves). Turn on the microwave long enough to boil the water. If the empty container heats up considerably as the water comes to a boil, it is not suited for microwave use. Some plastics will get so hot that they melt, whereas others (such as Styrofoam) remain quite cool. More heat being absorbed by the container means less heat going into the food. Many dishes absorb 15 to 20 percent of microwave energy; plastic wrap soaks up to 1 to 2 percent. But all dishes if left in long enough (ten to twenty minutes) will get hot. It's the ones that get hot in only two to three minutes that are not suited for microwave ovens.

Test Your Cooking Power

Microwaves are produced by an electronic device called a magnetron. The magnetron in a full-sized oven produces about seven hundred watts of power. Most compact ovens produce about five hundred watts. It's been shown that when baking cakes power levels below six hundred watts produce a wet, sticky crust. This is because the heating power is not strong enough to offset evaporative cooling on the surface of the cake.

Over time, many ovens experience a loss in power as the machine ages and internal power settings change. But there is a method for testing the power output of a microwave oven. Fill a pint container with water (two cups) and measure the water temperature. Heat the water in the oven for one minute and record the water temperature again. Multiply the difference in temperature (degrees Fahrenheit) by 17.5. The result is the approximate power generated by the oven in one hour. Theoretically it should match the manufacturer's rating, but so many factors are involved—hardness of water, container material, etc.—that this is just a ball park estimate. In general, if your microwave doesn't boil a cup of water in 1½ to two minutes, it's not working up to par.

Is My Microwave Oven Dangerous?

The real danger from microwave ovens comes from the possibility of microwave leakage. Door seals are the most important safety element to consider. Grease building up in the door seal attracts heat, which may weaken and warp the door. This can happen if the oven—and the seals—is not cleaned frequently. Electronic instruments specially designed for such purposes can measure door leakage, but a crude and harmless test can be made with your thumb. Run your thumb along the seal while the oven is on. If your skin feels warm, it's time to call in a professional to measure the leakage.

The glass on the oven door does not keep the microwaves in. The door screen behind the glass does. The screen is a piece of metal punctured with tiny holes. The microwaves are too large to go through the holes. The glass merely acts as a barrier to keep children from poking objects such as hairpins through the holes, causing electrical arcing across the points.

. . . .

TRIVIA: Don't cook tough cuts of meat in the microwave. The rapid heating does not allow the tenderizing process—denaturing—that occurs in slow cooking. Tough meat will just be made tougher. One alternative is to cook with moist heat or to use a very low setting on the oven. But in that case one might as well use a conventional oven.

Inventive cooks use their microwave ovens for more

than just cooking. On a rainy morning, the oven can be used to dry out the soggy newspaper quickly, and without fear of the paper catching fire. Paper is not a great absorber of microwaves. So water in the paper boils away while paper remains cool, i.e., below its combustion temperature. When paper is dry, it is cool enough to be handled.

. . . .

Done to a Turn

There are lots of ways to know when a food is done cooking: Probe it with a knife, poke it with a toothpick, listen for the timer to ring. For many foods, it's "done" means it has changed color. A well-done steak is brown. Light pink means rare. French fries are done when they turn golden or dark brown.

We take it for granted that heating food changes its color: The more you heat it, the darker it gets. But why does food change color when it's cooked?

It is not the flame or hot oil, not the heat itself that causes food to change color. Rather it's the chemical reactions *inside* the food, caused by high temperatures, that change food from raw to charred.

Consider browning: frying in hot oil or broiling. Browning occurs when parts of protein molecules (amines) combine with sugar at high temperatures. Both sugar and protein are necessary for this chemical reaction. That's why some fast-food eateries will add sugar to their french fries—to give them a golden brown color—and why on many barbecue sauce recipes you'll find brown sugar or honey as an ingredient. The sugar-amine reaction gives the barbecued meat a golden brown color. The sugar in roasting marinades does the same thing. Your meat is not brown enough? Just add more sugar. Does your recipe call for continuous basting? Constant basting keeps the sugar on the meat, where it can brown most efficiently.

Illustration 51

When experimenting with your browning recipes, be aware that table sugar—sucrose—will not work alone. By itself it will not react with the meat. It needs the presence of an acid. So be sure to include

132

vinegar or lemon juice in the recipe, or use brown sugar or honey. Vinegar and sugar are the key ingredients in many bottled sauces or cookbook recipes.

Just because a recipe may not call for sugar doesn't mean it's there all the same—as an ingredient in ketchup or in other processed foods. In his book *The Cookbook Decoder,* Art Grosser concocts a recipe for peanut butter barbecue sauce that requires lemon juice to activate the sugar in the peanut butter.

Color Me Pink

So much for the outside of the meat.

Doneness on the inside is a different story. Raw meat is red due to the presence of a pigment called myoglobin. In the living animal, myoglobin takes oxygen from the blood and transfers it to the muscles.

A truly fresh cut piece of beef is a dark, purplish-red, the color the butcher sees when he slices the animal open. This color is a good clue to freshness. But once meat sits exposed to the air, it will turn bright red as the myoglobin combines with oxygen in the air to become oxymyoglobin. Another way to produce the bright red color in fresh meat is to heat it. In this way the myoglobin combines with oxygen, too, and this is how rare beef gets its bright red color. But the red is transitory because if heating continues, the pigment disintegrates and forms a brown substance characteristic of well-done meat. Experienced cooks can usually tell from the color how much more cooking time the steak needs.

Knowing when a turkey is finished roasting takes much greater skill. Poultry, veal, lamb, and pork do not change color dramatically because they start out as light-colored raw meat. There is not enough pigment to use meat color as a reliable indicator. That's why good cooks always use a meat thermometer for doneness and why they always figure in the amount of additional cooking the roast will go through—standing time—as it sits and cools before eating (see "Nuke It in the Microwave").

TRIVIA: Some potatoes are not suitable for french frying because they have too high a sugar content. The sugar speeds up the browning process. Waxy potatoes, loaded with natural sugar, brown too quickly and are overdone on the outside before the potato has finished cooking on the inside.

Wrapping a turkey in foil slows cooking time. The metal foil acts as a heat shield, conducting the heat away from the roast so the bird cooks at a lower temperature and with less crust than an unwrapped turkey. The foil wrapping also locks in moisture, causing the bird to boil rather than roast. That's why many recipes call for a foil "tent": to prevent burned breasts and drumsticks without restricting the escape of moisture.

Cloud Seeding Your Drink

Tiny bubbles in the wine,
Make me feel happy, make me feel fine.
—As sung by Don Ho

No one knows when the first carbonated beverages were invented. One beverage company claims its product Perrier (bottled carbonated water pumped out of the ground) is "the earth's first soft drink." Claims like that are hard to verify, let alone dispute. Recipes found on clay tablets show the Babylonians making fermented alcoholic drinks six thousand years ago. But it's safe to say that carbonated beverages are very popular. In addition to natural spring water, carbonated beverages like beer, champagne, wine, and soft drinks are a multibillion-dollar industry. People like to feel the bubbles tickle their nose and carbonation adds a pleasing taste and texture that uncarbonated drinks lack. A glass of carbonated drink can be a wonderful place to explore the nature of bubbles.

Putting the Pop in Soda

Understanding the nature of fizz requires first understanding how the bubbles get into the drink. There are two methods, both involving infused carbon dioxide (CO_2). Soft drinks are made by forcing CO_2 into the liquid under pressure. The carbonation in beer and champagne is a product of fermentation—yeast breaking down organic material. In both types of carbonation, the gas is kept in the liquid by pressurizing the container.

When the bottle or can is opened, the pressure drops dramatically as gas escapes. With no pressure to hold them back, the bubbles in the drink form and fizz to the surface. This is the scenario that occurs every time you open a carbonated beverage. But it doesn't have to happen at all. Theoretically speaking, there is no reason why any fizz has to emerge from the drink.

A chemist might look at the drink and call it a supersaturated system. Unopened, the can contains the maximum amount of CO_2 the liquid can hold under that amount of pressure. There is no reason why the liquid, if not disturbed, cannot hold on to the gas and stay supersaturated forever, at least in principle. But life is not a perfect theoretical system and a few irregularities will break up the delicate smoothness of the system and cause the gas to bubble out. Such an irregularity is the act of popping the cork, twisting the cap, or flipping the pop top. These events set into motion the release of the bubbles. The gas rushing out disturbs the system, stirs up the liquid, and sets off the rapid formation of fizz.

Once the ruckus settles down, it can be started up again very easily. It takes nothing more than the act of pouring the drink over ice to set off the fizz again. Most people believe it's the cold cubes that release the gas, but that's not the case. Irregularities in the ice—little microscope nonuniform shapes and little pockets of air—serve as places where bubbles can form. Bubbles use these nuclei to grow. Even when there is no ice in a glass, the bubbles will find little irregular places in the glass itself. The glass may look smooth, but it has tiny microscopic pits and cracks that serve to create bubbles through the process of nucleation.

Bubble Streamers

Consider one of the most intriguing events occurring in a glass of carbonated drink: the familiar stream of bubbles that rises to the top from a fixed point on the side of the glass. Usually there is not one but many points, each with its own tower of bubbles leading to the surface. Where are those bubbles coming from? How can so many bubbles come from one tiny spot? Why do they continue to form? The spot that serves as the jumping-off point for the stream of bubbles is actually a tiny nick in the smooth glass surface. As the drink is being poured, a bubble of air gets trapped in this crevice, as shown in Illustration 52.

The air pocket begins to attract molecules of carbon dioxide, which leave the liquid and stick to the air. The molecules of CO_2 are caught in a tug-of-war. They are attracted both to the liquid and to the air pocket. But in the end the attraction of the air is stronger, so

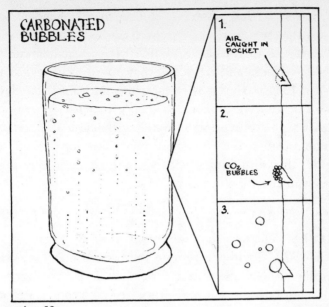

Illustration 52

Bubbles streaming up the side of a glass start out as tiny pockets of trapped air. Swelled by carbonation, the bubbles grow larger until they are buoyant enough to break loose and stream to the top.

they pile up along the surface of the air pocket. As more molecules join in, a bubble grows. The growing bubble soon becomes too buoyant to be held down, so it breaks away and rises to the surface. The whole process is repeated, resulting in a steady stream of bubbles.

(This same event occurs in a pot of boiling water. Steam bubbles form in crevices in the bottom or sides of the pot or kettle. When they get big enough, they float to the surface and another bubble forms in the crevice. That's why a pot of boiling water looks—and behaves— like a glass of carbonated drink.)

Cloud Seeding Your Drink

I once observed a beer drinker do something very unusual to his beer. Taking a salt shaker, the man shook a bit of salt into the glass. As if poured anew, the beer exploded in foam, and a drink that had been flat developed a new head. The man had no idea why it worked,

137

only that it was a trick passed on to him by a bartender. Soft-drink lovers know the trick works equally well with a carbonated soft drink, and its secret lies in the salt crystals. Salt crystals contain so many irregularities—cracks and crevices—that pouring them into a carbonated drink is like seeding it with locations for bubbles to develop. In fact, the situation is analogous in theory to what happens in cloud seeding. Air can become supersaturated with water and needs only the right kind of push to start the formation of raindrops. When scientists seed a cloud, they drop bits of salt, hoping that the salt's irregularities will serve as nuclei for the formation of drops of water.

• •

TRIVIA: The first soda water was made by Joseph Priestley in 1772. Trying to imitate the natural bubbling water of mineral springs water, Priestley added an artificial mineral—soda—to his new drink. Today's soda water may or may not contain soda (sodium bicarbonate). Seltzer originated in Selter, a town near Wiesbaden, Germany. Water from Selter came from an effervescent natural spring of high mineral content, but modern seltzer is produced artificially and is valued for its sodium-free quality. Although inventing soda water may have been refreshing, Priestley is better known (and respected) for discovering oxygen.

• •

V. Washing Up

Cleanliness is next to godliness—
and next to impossible.
—Anon.

A Real Soap Opera:
Making Water Wetter

Years ago, a TV commercial for a detergent touted its cleaning magic by claiming the product made water "wetter." The commercial went on to explain that while we all think of water as being naturally wet, what we really mean when we say water is "wet" is its ability to latch on to anything it touches: clothing, skin, dishes, or car finishes. In the laundry, garage, kitchen, or bathroom, this can be translated as the ability of water to stick to grease and dirt and "float it away." As anyone who has tried to wash a greasy face or a dirty dish knows, water alone does not work. It does not stick to the stuff and wash it away, it just beads up and

Illustration 53

rolls off. What the soap and detergent do to make water wetter—enhance its stickiness—is a fascinating bit of chemical magic with a long and interesting history.

Soap has been used for countless centuries without anyone really knowing why it worked. Soap was (and still is) produced by taking animal fat (tallow) and treating it with soda ash (lye) to form a compound that was later discovered to be sodium stearate. Sodium stearate belongs to a family of chemicals that are elegantly suited to overcome the incompatibility of oil and water. It's no secret that oil and water want to have nothing to do with each other. Thrown together, each one's natural inclination is to avoid the other, close ranks, and form into a big drop. But if you're going to get clothes, dishes, or skin clean, you've got to find a way to make water "wet" enough so it will overcome its integrity as a substance, stick to oil, and wash it away. And that's what soaps and detergents can do.

How? First, recall that water has a polar personality (see Chapter I, "By the Beautiful Sea). One end (hydrogen) is positively charged; at the other end (oxygen) is negatively charged. Oil, on the other hand, has an electrical distribution that is quite uniform—it has no poles— explaining why oil and water don't mix. What is needed to bring the two together is an intermediary, a substance that has the properties of both oil and water, partly polar and partly nonpolar. That's exactly how soap works.

The soap molecule looks like a long snake (see Illustration 54). One end is composed of a water-soluble sodium group, the other of a water-insoluble stearate chain. It helps to visualize the soap molecule as having a sodium "head" and a fatty "tail." When soap is added to water, it does not dissolve. The fatty tails repel the water and will do anything they can to get out. They rush to the surface and align themselves tails up in the air or into the sides of the sink. In other words, the soap molecules gather into small spherical bodies, with oil tails pointed toward the center and heads pointed outward.

When greasy or soiled fabric is added to the soapy water, or soapy water is put on your hands and face, the soap molecules rush to the newcomer (see Illustration 55). Soil usually has an oily film binding it to fabric or skin so the fatty tails of the soap find a compatible substance in the greasy mess. They rush to the oil and wedge their tails into it, trying to bond. Acting like millions of microscopic air hammers, they chip away at the large oil and dirt particles and separate them from the fabric. Freed from the fabric, the released

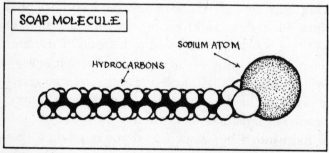

Illustration 54
Stylized soap molecule, with head and tail.

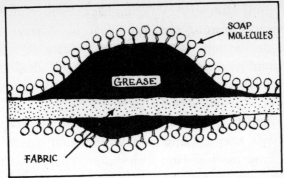

Illustration 55
Detergent circling grease with heads and tails shown.

grease coils up into small oil droplets that are again set upon by soap molecules that firmly imbed their tails like toothpicks in an olive. With their heads sticking into the water and their tails surrounding the dirt and grease, the soil-soaked soap follows the water down the drain. The layer of soap prevents the grease from being reimbedded in the fabric.

After World War II, America's love affair with washing machines opened a whole new field of washing chemistry. As more and more people gave up laundry soap and elbow grease for the modern washing machine, consumers were demanding better washing products that would make clothes clean in hard water. So chemists had to invent products more suitable for the times. They realized that they could keep the soap molecule as a model—the water-soluble head, fatty-tail structure—but they changed and added ingredients to it to create a whole new family of washing agents that were superior to plain soap: detergents. (Synthetic detergents were developed in the late thirties and early forties but were widely used only after World War II.)

Read the ingredients on a box of today's laundry detergent and you'll find a laundry list of additives. Topping the list are water softeners. Detergents and soaps will not lather (foam) very well in hard water. Hard water is loaded with calcium or magnesium in the form of free-floating, charged particles called ions. When soap meets the calcium ions, the first thing it does—before starting to clean—is to react with the calcium and precipitate a salty, gummy residue:

143

bathtub ring. Only after all the calcium has been precipitated does the soap get to work cleaning the dirt—that is, if there is any soap remaining to do the job. Water softeners replace the calcium with sodium. Sodium ions don't interfere with the cleansing process. Detergents became very popular mostly due to the water softeners that made them work so well without leaving a residue.

Water softeners in the form of phosphates were so widely used that in the 1960s they became an environmental hazard. The phosphates in washing machine and dishwasher water, as it poured into lakes, promoted the growth of algae and caused fish to die. On many of today's detergents you will find the words "contains no phosphorus" written in bold, capital letters.

Next on the list of ingredients, if the detergent is liquid, may be alcohol or propylene glycol to make the detergent pour. Soil-suspending agents help keep the soil from being redeposited on the clothing. Enzymes try to digest stains. Antifoaming agents keep the suds down in the washer; foaming agents encourage lather in shampoos.

Some detergents are advertised as making whites "whiter." If white is white, what does whiter mean? Many times a white fabric will yellow a bit with age, and while the garment is still perfectly usable, our culture (advertisers) has convinced us that yellowing is not desirable. So to sell more soap, detergent manufacturers first added "whiteners," many times nothing more than a blue coloring agent. Blue absorbs yellow light, causing the yellowed garment to appear white. After World War II it was found that "optical brighteners" would do a better job than coloring agents. Optical brighteners act by adding blue light, not blue dye. They absorb ultraviolet light, change its wavelength, and reemit it as blue light. Blue light mixes with the yellow and adds up to white. Brighteners do the job so well that sometimes you swear the bright pillowcases and towels have just a tiny hint of blue in them—which they do.

. . . .

TRIVIA: Detergents became so widely used after World War II that streams and rivers became loaded with suds. The detergents were not easily broken down by nature.

Once they left the washing machine, they continued to "foam and suds." The problem was solved by the use of "biodegradable" detergents; now just about every maker of detergent makes sure the word "biodegradable" is prominently displayed on the container.

Eggs do for salad dressing what soap does for dishwater: bring disagreeable ingredients together. In the same way that soap brings oil and water together, eggs bring oil and vinegar together. Egg yolk in mayonnaise contains lecithin, a natural emulsifier. Lecithin is built like soap in that one end of its molecule (tail) likes oil, while the other end (head) likes water. By surrounding the large oil globules in salad dressing, the emulsifier breaks up the droplets into smaller ones encircled by egg yolk. With the water-loving heads protruding, the emulsified oil is easily dissolved by the water or vinegar.

. . . .

The Shower As Concert Hall

Swans sing before they die—'twere no
   bad thing
Should certain persons die before
   they sing.
—Coleridge

Some people say it's the hot water. Others claim it's the solitude. But whatever the reason, there's no place like the shower for launching into a tune. They even make a shower radio for music diehards.

It's easy to see why we sound so much better in the shower: the shiny, bare walls. The sound is reflected many times by the smooth bathroom tiles, bouncing around the room, setting the room "ringing." This echo effect, the ability of the room to take our voices and "sing" with us, is what we call resonance. Resonance is an essential part of sound production, whether the sound is a closet coloratura singing in the bathroom or a symphony orchestra playing Mozart.

When an object resonates it is taking a sound produced by a weak source and amplifying it. Take the strings on a violin. If removed from the instrument, stretched, and plucked, the strings could hardly be heard. But when mounted on the resonating box of the violin, the strings produce rich tones that carry to the back of a concert hall. The same action occurs with our voices. We could sing to ourselves all day and no one (thankfully) might hear us. But put us in a bathroom, which resonates with the sounds of our voice, and we're all opera stars. In this case, resonance is produced not by the small air-filled violin box but by the large air-filled space of the bathroom set in motion by vibrating vocal cords.

Singing produces high and low notes. Each note is made of a sound wave that has a certain length. The higher the pitch, the shorter the length of the sound wave. The lower the pitch, the longer the

Illustration 56

wavelength. These sound waves bounce between the walls and rebound off the ceiling and floor, bumping into one another. Of all the notes bouncing around your bathroom or shower stall some are just the right length to fit neatly between the walls. As you sing, these notes align themselves between the walls (and floor and ceiling), literally piling atop one another right in the center of your shower stall. The net result is that these select notes sound louder than others; they stand out. They cause the bathroom to ring, to resonate; or the other way around, the bathroom resonates at the wavelengths of these notes.

Measure Your Shower

The rules of resonance for a bathroom situation say that the strongest resonance occurs when the distance between any two walls is equal to half the wavelength of the note being sung. Since the distance from floor to ceiling is different from the distance between opposite walls, the room resonates at at least two different frequencies. What are these frequencies? (Actually, the situation is slightly more complicated, but for illustration purposes let's simplify.)

Let's look at a typical shower situation with a shower stall about three feet wide. Its floor-to-ceiling height is about eight feet. Because there are two different lengths—width and height—there should be

147

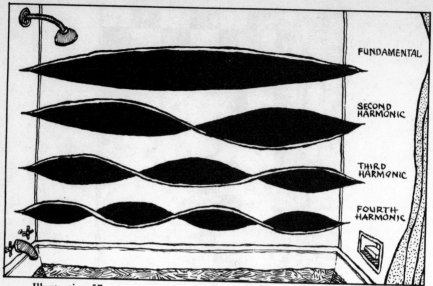

Illustration 57

Shower resonance. Sound waves vibrating between the walls of
a shower so that one, two, three, four, etc., half wavelengths
fit exactly in the space between the walls. The top wave is the
fundamental. The bottom three are the second, third, and
fourth harmonics. The fundamental wave is the tallest,
indicating it resonates the loudest.

two different resonances. According to the rules of resonance, the
wavelengths of its fundamental frequencies—the strongest resonating
note is called the fundamental—must be double these dimensions,
that is, sixteen feet and nine feet. The wavelength of a note is equal to
the speed of sound divided by the frequency of the note. The speed of
sound is standardized at about 1,125 feet per second. Taking 1,125
and dividing by sixteen and nine, respectively, yields wavelengths
corresponding to notes D and B, two octaves below middle C. These
notes will ring loudest in your shower stall. If you sing a little higher
than the resonant note, the sound dies down. Sing a little lower and
the same thing happens. But notice: These are low notes, below the
tenor range. Does this mean that only people with a bass voice sing
well in the shower? Obviously not. Harmonics play a key role in
resonance. Harmonics are multiples of the fundamental frequency, and

they also fit very well between the walls, as in Illustration 57. They cause resonance, too, though they don't produce as loud a note. So the shower can resonate at different frequencies, which explains why everyone finds it such a satisfying place to sing. (It's clear why the walls of concert halls have to be specially designed to reflect sound waves. You don't want the room to resonate like a shower stall.)

Next time you're soaping up, try singing a musical scale. At those times when it sounds like the walls of Jericho tumblin' down, that's when you've hit the fundamental resonant notes.

Breaking Wineglasses

Resonance can be very powerful. It's been said that great opera stars like Caruso could shatter wineglasses with their voices. No magic is involved. It's just a matter of finding the resonant frequency of the wineglass and singing loudly enough. At this frequency a wineglass will literally blow itself apart from the inside, succumbing to the force of the sound waves bouncing off its walls.

Any object that vibrates will have a resonant frequency. Squeaks and rattles in a car may not be very annoying unless the car is of just the right size to make the noise resonate loudly. What in one car would be a minor annoyance would be intolerable in another if the acoustics were right. On the other hand, sometimes resonance in a car can be very useful. Many people prefer to install speakers in the rear of their car because sound produced in the backseat is much fuller. The low frequencies find a nice place to resonate in the unhindered expanse of the automobile.

Resonance is a common occurrence in nature whenever sound vibrations are involved, and it has a similar effect in the world of radio waves. Radio waves bounce not between walls, but "vibrate" in antennas. To make a radio or television set work most efficiently, the length of the receiving antenna must be made to fit the wavelength of the incoming signal. If the wavelength of the signal doesn't match the antenna—if it's too big or too small—the antenna won't resonate well. It won't "vibrate" electrically as strongly as it would at the resonant frequency, so the signal is lost. Interestingly enough, the optimum length for antennas is exactly the same as the optimum length between walls in our bathroom: a half wavelength. If an antenna's length is half as long as the wavelength of the incoming signal, the signal will transfer most of its energy to the wire.

· ·

TRIVIA: The most dramatic demonstration of the power of resonance occurred on November 7, 1940, in Tacoma, Washington. A high wind set the Tacoma Narrows Bridge vibrating. The bridge, resonating with the wind, pitched up and down and swung out of control. Finally it tore itself apart. As a result, soldiers marching in unison across bridges are frequently told to break stride so as not to set up vibrations that may set off another violent resonance.

The sound of the "ocean" in a seashell is due to the shell's resonating effect. The sound that is heard is at the wavelength at which the air space in the seashell resonates. The shell acts as a resonator to magnify the external, ambient sounds that match the size of the shell's cavity. The shell produces no sound of its own.

· ·

Tornado in the Drain

There is something that goes on in the bathroom sink and bathtub that has kept scientists scratching their heads for decades. It is nothing of earth-shaking proportions. But it is a phenomenon that physicists are continually trying to explain to themselves. It's the question of why water swirls when it goes down the drain and in what direction that swirl should be: clockwise or counterclockwise.

Who hasn't pulled the plug on the sink or bathtub and watched the water form a little whirlpool as it flows out of sight? But why should this phenomenon occur? Why doesn't the water go straight down the drain without riding its merry-go-round?

For years the simple answer has been the *Coriolis force*. The Coriolis force (or effect) is responsible for the destructive swirls of hurricanes and tornados; it causes the circular movement of high- and low-pressure fronts. And some scientists believe it is the active force that makes the water swirl down the drain.

How does it work?

The Coriolis force is due to the eastward rotation of the earth. The rotation influences any moving body of air or water and causes it to be deflected to the right in the Northern Hemisphere and to the left in the Southern Hemisphere.

If a rocket ship were launched from the equator toward the North Pole, the rocket would also be moving eastward. At the equator, the earth is rotating eastward over a thousand miles per hour, and that speed is imparted to the rocket. But north of the equator, the speed of the earth's rotation falls off as the circumference of the earth gets progressively smaller. (It's akin to a phonograph record spinning faster on the outside than near the spindle.) At 30 degrees north latitude— Houston or Baton Rouge—the speed of the earth has dropped to about 935 miles per hour. But because our rocket ship is still moving independently at its initial higher speed, it will be drifting eastward

Illustration 58

relative to the earth's surface. So to someone standing in Texas or Louisiana the rocket will appear to be curving eastward. This is the Coriolis effect.

If the rocket had been aimed at the Empire State Building without taking the Coriolis effect into account, it would have missed the New York City landmark by a wide margin to the right as its path curved eastward under the influence of the effect. (The actual path taken by the rocket depends upon the observer. A man on the moon watching the event would see the rocket travel in a straight line, while people on the ground would see a curved path. Rocket ships and even airliners have to aim slightly left if they're to reach their destinations.)

A rocket ship fired from the North Pole toward the equator would experience the same effect and appear to curve toward the right, this time westward. The earth's slow polar rotation would give the rocket only a slight boost eastward, leaving the rest of the world below whizzing quickly by. You could no more hit Cleveland or Paris aiming directly from the north than you could if sighted from the equator.

In the Southern Hemisphere, the situation is exactly the oppo-

152

site: The Coriolis effect makes the path of objects curve toward the left.

Substituting moving air (the wind) or moving water (ocean currents) for our flying rocket leads to an interesting occurrence. The right-turning effect of the Coriolis force (in our hemisphere) makes any wind or water want to flow to the right. Major ocean currents, such as the Gulf Stream, are deflected as well as major movements of winds, such as weather patterns. The Coriolis effect explains why high-pressure weather fronts spin clockwise—the winds are deflected to the right.

The counterclockwise spin of low pressure fronts is also due to the Coriolis effect. Areas of low pressure act like holes, atmospheric drains that suck in high-pressure winds. As the winds are drawn to the areas of low pressure, their straight-line direction of travel is deflected to the right by the Coriolis effect. The combination of both forces results in the circular, counterclockwise swirl, of areas of low pressure such as hurricanes and tornados.

What does all this have to do with water running down the drain? If we consider the drain to be an area of low pressure—a hole— then water in our northerly tubs should, in theory, swirl down the drain like a minicyclone in a counterclockwise fashion. But this doesn't always happen. Pull your drain plug a few hundred times (a dozen, maybe?) and chances are if your drain is like mine, the water will randomly swirl in both directions.

Does this mean that the Coriolis effect is not working? This is the crux of the problem, a topic of debate among scientists. Many scientists believe the effect can be seen only on large bodies of water, perhaps those that are a quarter of a mile or more in diameter. That's a big bathtub. But others are convinced it can be observed even in the cramped quarters of your bathroom, but only under controlled, laboratory-like conditions.

These true believers are so convinced that they went to all the trouble of building special research bathtubs designed to do away with outside influences that might set the water inadvertently spinning in either direction. Their bathtubs were perfectly round and the sides and bottoms perfectly square (as opposed to the sloped floors on most tubs) so as not to give the water any initial movement before the pulling of the plug. After filling the bathtubs, they waited hours, even days, for the water to settle down so that any residual turbulence would die out. They controlled for any wind disturbances in the room and any other external influences they could think of; even the drain and plug were specially built.

Then they toted their tubs to locations north and south of the equator, filled them with water, and pulled the plugs. And sure enough, through many fillings and drainings, the tubs in the north drained counterclockwise, those in the south clockwise—or so claim the researchers.

What about right on the equator? A group of scientists found themselves staying in an African hotel on the equator. In a totally unscientific experiment, they filled the tub and pulled the plug and swore the water went straight down the drain without swirling in either direction.

(Many years ago, when I first tried these experiments on radio, a listener wrote that after hearing my reports he found himself aboard a U.S. Navy ship crossing the equator and at the exact moment of passage, he pulled the plug on a sink and he, too, watched the water head straight down the drain.)

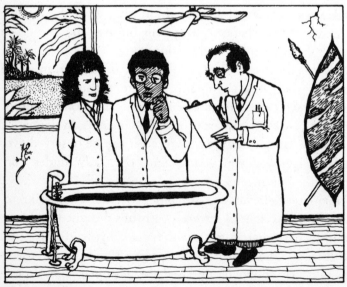

Illustration 59

Despite the research, many scientists remain unconvinced that the Coriolis effect has any noticeable influence on such a small scale. (Even though the toilet flushes in a vortex, the swirl is caused by jets of water cleaning the bowl.) So next time you're washing your face or draining the tub, conduct your own experiment in full knowledge that serious scientists are conducting theirs in bathtubs around the world.

VI. Aches, Pains, and Medical Myths

I'm not afraid to die.
I just don't want to be there
when it happens.
—Woody Allen

Why Men Grow Bald and Women Don't

If there is one truly vain streak in a man, chances are it involves his hair. Show me a man losing his hair and I'll show you one who worries more about that than he does his job security. Unfortunately, going bald is something no one can prevent. If it happens to you, there's nothing much to do. Some people try hair transplants; for others, it's a toupee or a shaved head. Why is it that only men suffer? Why are women lucky enough to go free?

The truth is women do go bald, but it's not as obvious. Instead of going completely bald, says Dr. Maria Hordinsky of the University of Minnesota, women will thin out on the top of their head and usually retain their frontal temporal hairline. Balding can affect men, women, and children, but the most common balding problem is "male pattern baldness." This problem gets its name from the balding pattern—loss of hair from the top and front, with the sides and back left intact. No one knows the true incidence for male pattern baldness, but simple observation will show that it is widespread. Lots of men—from Willard Scott to Carl Reiner—display the typical shining pate.

Illustration 60

Baldness involves a disruption in the normal hair growth cycle. Hair grows in four stages. In the first stage, new hair is initiated at the dermal papilla, the root of the developing hair. As the hair grows upward it eventually pushes out the old hair. As the hair develops, the papilla reaches down deeper into the skin and begins to grow into a nice healthy long hair shaft firmly attached to the scalp. This is the

159

hair we all like to have, of which TV commercials are made: long, luxurious, shining. About 80 percent of the hair on the scalp is in this deeply rooted stage.

The hair cycle continues, passing next into the transition stage. The hair is smaller, and it's loosely held to the inner part of the follicle. About 1 or 2 percent of our hair is in this stage of development. Finally comes the last stage of the hair cycle, the point when the hair falls out. Just a slight tug is all that's needed.

People normally lose twenty-five to 125 hairs a day. These are the hairs that come off on the comb, clog up the sink or shower drain, and latch on to your clothing (or someone else's). This loss is nothing to worry about; it's part of the cycle. New hair is growing all the time.

Most of the time, the cycle repeats itself with the dermal papilla supplying a new hair to take the lost hair's place. But in male pattern baldness, instead of the hair cycle starting over again, no new hair shaft replaces the lost one. The papilla initiates the growth of a new hair shaft, but the hair does not grow to maturity. Its growth is arrested in the "peach fuzz" state, and the heads of men with male pattern baldness may exhibit this short, soft hair.

When searching for clues to the cause of male pattern baldness, dermatologists used to theorize that hair loss was inherited from the mother's side of the family. No one believes this anymore. Dermatologists now think that hereditary baldness is a polygenic trait—that is, that many factors are involved. Perhaps a hormone in the skin or an environmental stimulus triggers the balding cycle. You simply can't blame it on your mother or grandmother.

The hair-growing process is not an easy system to study. In addition to the hair follicles in the scalp, there are connective tissue, multiple skin layers, and sweat glands that complicate matters. It would be nice if someone could find a way of growing hair in a test tube. Of course, then we would know how to make hair grow and no one would need to grow bald ever again.

The most promising, clinically tested hair stimulant is a substance produced by the Upjohn Company called Minoxidil. Minoxidil is a blood pressure medication, normally swallowed in pill form. But researchers discovered that in about 30 percent of balding people, the drug reawakens the dormant peach fuzz hair shafts and stimulates them to grow to maturity. Studies show that it is not the panacea everyone hoped it would be—it does not bring back long, lush hair to everyone—but it does stop hair loss in about one out of three people who use it.

Alcoholic Chaos

People who insist upon drinking
before driving, are putting the
quart before the hearse.
—G. K. Chesterton

Alcoholic beverages have been consumed since the beginning of history; the presence of beer and wine is well documented in archaeological records of the oldest civilizations. Yet few drinkers really understand alcohol's actions on the body. Myths, misconceptions, and rationalizations cloud most people's ideas about the effects of the beverage.

You've probably heard one or more of the following comments:

- "I actually drive better after a couple of drinks. I'm more relaxed."
- "She's not an alcoholic. She only drinks beer, not the hard stuff."
- "Have a cup of coffee for the road. It'll straighten you out."

They all reflect misunderstandings and wishful thinking about the profound effect alcohol has on the body. Alcohol profoundly affects every organ. The damaging effects of chronic drinking read like a list from a med student's homework assignment: muscular disorders, birth defects, digestive abnormalities, cancer, diabetes—the list goes on and on. It's beyond the scope of this book to detail each one. But one observation is clear: We may put alcohol into our stomachs, but it finds its way into every part of our body.

The kind of alcohol that is drinkable—the active ingredient in wine, beer, or spirits—is called ethyl alcohol, or ethanol. A molecule of ethanol is made of two carbon atoms, a ring of hydrogen atoms, and

Illustration 61

one oxygen atom. The exact makeup is not important. What is important is that the molecule is very small; it can go anyplace it wants to, and when it gets there it makes its presence felt.

Immediate Effects

Eating and Drinking

On the way to the stomach, the tiny molecules of alcohol interact with the throat cells creating a burning sensation, a warning of the chaos to come. How fast alcohol gets into the blood depends in part on the speed in which it reaches the small intestine. Most of the alcohol is absorbed in the first sections of the small intestine, so anything that causes alcohol to remain in the stomach will actually retard absorption. A concentrated drink will irritate the lining of the stomach and may paralyze the muscles of the stomach wall, causing the pylorus (the valve between the stomach and the small intestine) to spasm (close up). When the opening to the small intestine is closed, the drink will stay in the stomach longer, slowing its absorption into the blood. Mixers and carbonated drinks like sparkling wine, champagne, or club soda tend to relax the pylorus, speeding the passage of the alcohol into the small intestine. The amount of mixer used in a drink also influences the rate at which the alcohol is absorbed into the blood. The more dilute the drink, the slower the absorption rate.

Eating just before or during drinking slows absorption of alcohol, especially if the foods are rich in oils or milk products. A heavy meal may slow absorption so much that the highest blood alcohol concentration (BAC) may not be reached for six hours. More importantly, eating may lower peak blood alcohol concentration by slowing the absorption rate. Low concentrations of alcohol stimulate secretions of gastric (stomach) juices, which explains why a glass of wine before a

163

meal is reputed to stimulate the appetite. On the other hand, high concentrations not only irritate the stomach lining but also inhibit the digestive enzymes.

Burning It Up: Next Stop, the Liver

Almost as soon as alcohol enters the bloodstream, the body begins to get rid of it. A small percentage is given off unchanged in urine, perspiration, and breathing. But the bulk is "burned up" (oxidized) by the liver. The liver is a large, complex organ located in the middle of the body. All food, drugs, and chemicals—anything that's in the blood—is processed by the liver. One of the functions of the liver is to detoxify poisonous substances like alcohol. About 75 percent of the oxidation of alcohol takes place in the liver. The rate at which metabolism of alcohol occurs is somewhat variable, but in general a healthy 150-pound man will usually metabolize a standard drink (0.6 ounce) in roughly two hours.

After just one drink the alcohol moves through the liver and into the bloodstream, and heads right for the brain. The brain lets you know when the alcohol has arrived. First to feel the effects is the largest area of the brain, the cortex. Thought processes are centered in the cortex, so the first sign of the arrival of alcohol is muddled thinking. In many people, the muddledness can set in just ten minutes after the first drink. At a BAC of about 0.05 percent, judgment and restraint may be more lax; ordinary anxieties and inhibitions are diminished.

As the alcohol level in the blood increases, a depressant action kicks in. The alcohol then disrupts the second area of the brain—the cerebellum. The cerebellum is in charge of motor coordination, so an invasion by alcohol disrupts normal walking or driving, as well as reading and writing. At 0.10 percent—in many states the legal limit of alcohol in the bloodstream—voluntary motor actions become clumsy and reaction time slows.

At 0.20 percent, the entire motor area of the brain becomes significantly depressed; emotional behavior also becomes affected. People stagger and lie down; they may easily be angered, shout, or cry. At a BAC of 0.30 percent, the more primitive perceptive areas of the brain are dulled: People become confused, stuporous.

Finally, imbibe enough and the alcohol reaches the part of the brain controlling consciousness and breathing. At the 0.40 percent mark an individual may lapse into a coma. A BAC of 0.50 percent depresses the medulla and usually causes death by respiratory failure within a couple of hours.

Note that alcoholic intoxication is grouped by BAC—level of alcohol in the blood—not by the number of drinks consumed. A given amount of alcohol will produce different blood alcohol levels and thus different behavior effects in different people. For example, fatty tissue contains less water than muscle tissue. Because women generally contain more adipose (fatty) tissue than men,

Illustration 62

and because alcohol tends to dissolve in the water of tissues, a given amount of alcohol will remain more concentrated in the fluids of a woman's body than in a man's of the same weight, causing her to be more intoxicated.

Contrary to the widespread notion that people used to driving while under the influence of alcohol are unimpaired by it, this is not the case. Tracking and coordination functions of the eyes become progressively impaired as the blood alcohol level rises. Regardless of whether a person feels more relaxed or confident in his or her driving, the erosion in skill, judgment, and reaction time that alcohol produces places the driver at higher risk of making a mistake in a critical situation and having an accident.

Holiday Heart Syndrome

Low to moderate drinking will produce a slight, fleeting increase in heart rate and blood pressure. Blood vessels at the surface of the skin dilate, letting body heat escape and producing a flushing or reddening of the skin. People mistakenly believe that drinking before going out into cold weather will keep them warm. But just the opposite effect occurs. An artificial sense of warmth results in a misplaced feeling of security which, in many cases, can lead to death for people who did not bundle up enough.

Heavy drinking may reduce the pumping power of the heart and

Illustration 63

disturb heart rhythm. Brief drinking sprees in apparently healthy people have also been shown to produce disruptions in the heartbeat. This disturbance has been named "holiday heart syndrome" because of the increased incidence of irregular heart rhythms observed during holiday seasons.

Drug Interaction

Alcohol has become such an accepted part of our social functions that people who take medication often forget that they are already taking a drug: alcohol. Regardless of whether the other drug is a "strong" illegal one like cocaine, or a "mild" over-the-counter cough medicine, combining it with drinking is potentially dangerous. When mixing alcohol with drugs, the addition of one plus one may add up to more than two. A case in point is mixing barbiturates (sleeping pills) or Methaqualone (Quaaludes) with alcohol. Either one taken with just a small amount of alcohol causes an unexpectedly severe depression of the nervous system. A blood alcohol level of just 0.10 percent has been known to cause death when barbiturates were also present.

If a drug acts additively with alcohol, as happens with antihistamines, the effects—in this case, drowsiness or sedation—become intensified beyond the level either drug would produce if taken alone.

At the other extreme, if an antagonistic interaction takes place between the two drugs, the alcohol could cause a necessary prescription to be partially or totally ineffective.

Long-Term Effects

Liver Damage

As can be expected from its central role in detoxifying alcohol, the liver takes the brunt of abuse in chronic drinking. Liver damage occurs in three stages: fatty liver, hepatitis, and cirrhosis.

In the first stage, the liver begins to get greasy and full of fat. Eight drinks a day—wine, whiskey, or beer—for two or three years is enough to turn the liver a fatty, yellow color. There are usually no external signs of this problem. The liver still functions at this point, and if chronic drinking is stopped, the liver will heal itself and regain its rich, healthy color.

But if heavy drinking continues, hepatitis—inflammation of the liver—may develop. In this stage the liver is enlarged and tender. Jaundice is usually present. Areas of cell death occur. Complications leading to death may set in.

Keep the liver bathed in alcohol over a period of ten years and the liver is overwhelmed. It becomes very hard, very puffy. The healthy tissue has been replaced by scar tissue produced by the damaged liver. The liver is virtually shut down and doesn't work very well if at all. This is called cirrhosis and it means big trouble. The scar tissue does not function. It cannot be replaced, nor can the damage be reversed. Continued drinking at this point surely leads to death, but many people die of complications before advanced cirrhosis is reached.

Nervous System

Just as the tissue in the liver was destroyed by the alcohol, so too the cells in the brain are poisoned. Alcohol can enter the membranes of the cells, making them more fluid and destroying them in the process. Various disorders of the central nervous system have been diagnosed in 50 to 70 percent of alcoholics entering treatment.

Sludge Formation

There is some evidence of the possibility of damage to the circulatory system in long-term drinkers. This damage is called sludge formation. Alcohol may cause red blood cells to clump, slowing circulation—possibly stopping it—in very small blood vessels. The evidence for sludge formation is not conclusive. But if this does happen, it could cause damage to the nervous system, heart, and muscles, by depriving cells of oxygen. Chronic consumption can lead to heart failure through a direct toxic effect on heart muscle.

Feeling Good

Why, then, if all this torture is being inflicted on the body, do people continue to drink? The answer is, in part, obvious. Drinking

167

makes people "feel good," and there is a biochemical reason for that good feeling. Inside the brain are opiatelike substances called endorphins, the brain's natural pain-killers. Alcohol interacts with the endorphins and binds them in an area deep inside the brain called the hypothalamus. This interaction is perceived as a very joyful experience, and people who become addicted to alcohol come to rely on this positive sensation as a substitute for their normal good feelings. Why we all don't become alcoholics may have to do with a genetic component to the disease. It is thought that some people are more prone to the addicting euphoric qualities of alcohol in the hypothalamus. This trait may be passed from one generation to another and might explain why some people cannot leave the bottle alone while others can take it or leave it.

• •

TRIVIA: A drink of wine or beer has the same amount of alcohol as a drink of "hard liquor." Alcohol is less concentrated in beer or wine than in spirits, but the amount of alcohol in a typical serving of each is about the same: a twelve-ounce can of beer, a five-ounce glass of wine, and a shot of eighty-proof whiskey all contain the same amount of alcohol, a little more than half an ounce.

• •

Don't Suck on a Snake Bite

Illustration 64

Just about every Boy Scout of my generation was taught what to do for a snakebite: Cut open the wound and suck out the venom. The Boy Scout handbook had detailed instructions showing how to lance the bite marks with a sharp knife and how to extract the venom using one's mouth or the suction cup tucked away for such emergencies in the first-aid kit (Be Prepared!). None of the Scouts I knew had ever been bitten by a snake, so there was no way of knowing from personal experience if the snakebite first-aid technique worked. That's just as well, because the most recent research concerning the treatment of snakebites now repudiates that timeworn advice. The experts now tell us not to suck out the venom because it may do more harm than good.

The damage done by a snakebite depends on three factors: where you're bitten, what type of snake bit you, and how much venom is injected at the time of the bite. The extremities—the hands, arms, legs, or feet—are the most common places to be bitten and are the least life-threatening. A snakebite in the center of the body—the upper shoulders, neck, and head—are naturally higher-risk areas.

Venom disrupts cell membranes and digests cell tissues. It begins working immediately. Badly damaged tissues do not recover. They die, slough off, or are surgically removed if gangrenous. The toxicity of the venom depends upon the kind of snake. Rattlesnake venom has a complex combination of proteins that affect the circulatory system. Rattler venom will influence how blood clots, flows, and drains into different tissues. Cobra venom has some of the same properties of the rattler, but more importantly, cobra venom attacks the nervous system. It can paralyze muscles controlling breathing or heartbeat.

Illustration 65

Of course, the amount of venom will influence the intensity of the poisoning. Amazingly, in 25 to 50 percent of all snakebites, no venom is injected into the bite. The controlling factor is the motivation of the snake. If it's looking for a meal, the snake will inject a lot more venom into its victim than if it is striking out of self-defense or fear.

As a rule, snakes don't bite and hang on like bulldogs. Snakes bite very rapidly in a stabbing motion. The whole process from strike to injection takes about one one-hundredth of a second, regardless of whether venom is injected or not.

Of course, there are definite first-aid steps one should follow after being bitten, starting with a list of "don'ts." First, do *not* do the old cut-and-suction trick. A deep, jagged incision made by victims playing amateur doctor creates a dangerous wound open to infection. As for removing the venom by suction, toxicologist Dr. Dan Keyler says it has yet to be proven that sucking on a snakebite will remove enough venom to make the whole cut-and-suck effort worthwhile. The maximum amount of venom removed from a bite by trained specialists is about 18 percent. If you had a very venomous, potentially fatal bite, that small amount might be the difference between life or death, but chances are the snakebite is not that life-threatening. Your crude incision may cause more harm than good.

Second, don't put ice on the bite and don't use a tourniquet to stop the bleeding. The old ideas about stopping circulation to the area are no longer accepted. In place of a tourniquet, what's now used is a constriction bandage—much like an Ace bandage used to wrap a sprained ankle and wrapped with about the same degree of tightness above the bite. The bandage retards the flow of venom carrying blood, slowing its movement up through the main part of the body. Third, don't give antivenom

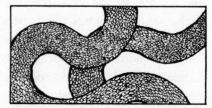

Illustration 66

170

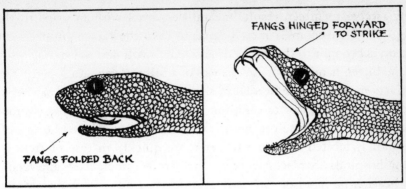

Illustration 67

medication to yourself or a companion in the field. Wait until you get help at a medical facility. Antivenom medication should be delivered intravenously and by trained medical personnel. It has to be carefully measured and diluted. There is always danger of shock or allergic reaction resulting from the medication that you may not be able to cope with. Only in extreme situations should self-medication be attempted. For example, in Third World countries, people routinely inject themselves with antivenom medication. In Burma, bites from Russell's viper are the fifth leading cause of death. In such a country in which the victims are many days from help, people have learned to carry antivenom snake medication and to inject the drugs into the muscles. Out in the bush this is the only chance for survival, since a shot in the arm or leg is better than none at all.

The experts now say that the single most important thing you can do is to keep the victim calm. Have him or her lie down and relax. This may not be easy to do, as anyone who has tried to calm a hysterical person knows, but it's the most crucial advice. Once down, the victim's bitten extremity—be it an arm or a leg—should be slightly bent and placed at heart level, not above the chest or below the chest. The object here is to prevent a change in the circulation to the limb—to prevent a decrease or increase in circulation. Speeding up circulation increases the rate at which the poison reaches vital tissues. Slowing down circulation or cutting it off by use of a tourniquet merely deprives the limb of much-needed blood.

Chances of surviving a venomous snake bite are excellent. Of about eight thousand venomous bites per year in the United States,

171

only twelve to fifteen result in fatalities, even without trips to the hospital. Antivenom is mostly used to save the bitten limb. If the victim is brought to the hospital, antivenom will neutralize the poison in the bitten hand or foot, so chances of a full recovery are superb. The antivenom may not be needed to save your life but it could save your limb. Therefore, prior to camping or hiking in potentially dangerous and remote locations, the wise traveler finds out where the nearest hospital is located and how to get there quickly in case of trouble. Most hospitals in places where snakes are common have antivenom medication on hand or can reach it quickly. And it's not uncommon for hospitals to dispatch planes to other parts of the world to fetch antivenom to treat a bite from a rare snake.

• •

TRIVIA: Horses help to make snakebite medication. Small doses of snake venom are injected into horses. The horse's immune system is roused to fight off the venom and thereby produces antibodies to the poison. The horse blood is then drawn off and the antibodies are purified and used to make antivenom.

• •

VII. Armchair Astronomy

Look not thou down but up! . . .
—Robert Browning

When I was a boy, my friends and I used to lie on our backs and gaze up at the stars. The hood of a car, a patch of grass, or a comfortable chaise lounge was just the ticket for stargazing on warm summer nights. Our childhood conversation of baseball and horror movies would come to an abrupt stop at the glimpse of a tiny speck of light moving slowly but ever so distinctly across the sky. This sighting of a manmade satellite, be it *Echo I* or *Sputnik I,* was enough to make a ten-year-old believe that maybe someday one of us might be riding a spaceship to Mars or the moon.

Today, space travel is so commonplace that my childhood miracles—satellites—are taken for granted. Newspapers no longer carry a daily schedule of satellite viewing times. But the lure of the heavens has not been diminished one bit for us diehard romantics. The stars and the planets still hold us in their grasp. Hardly a night goes by when an eye is not cast to the sky, perchance to see brightly shining Venus, or to be treated to a breathtaking glimpse of Jupiter or Saturn.

Even now, despite aging footprints on the moon and close-up photos of Halley's comet, outer space still has the power to mesmerize us. Viewed through the eyes of binoculars, the moon has lost none of its splendor. The discovery of black holes and pulsars add new mysteries never dreamed of in our youth.

The musings on the following pages can hardly do justice to the grandeur of our cosmic neighbors, but if when finished with this chapter you find yourself eagerly anticipating the coming of nightfall, you've not been bitten by a vampire but by a curiosity and wonder that's as old as astronomy itself.

Illustration 68

A Note on Night Vision

Whenever someone asks whether I dream in color, I have to sit back and think about it. I'm never quite sure.

It's the same thing with night vision. If I asked you whether you see colors at night, what would you say? Think about it. Do the leaves on the trees look green in the darkness of your street? How about the grass—what color is it? The point is, our eyes lose their ability to see colors in darkness. Objects we think we are seeing in color are really being seen in black and white and shades of gray like the images on old TV sets. Why?

Our eyes have two sets of sensory organs—cone-shaped and rod-shaped. The cones are sensitive to color but not to weak light. The rods can't tell one color from another but they can "see" extremely weak light. So as the sun goes down and the light fades, the cones lose their ability to see color but the rods keep on working.

You probably won't notice the loss of color in night vision unless you look for it; our brain tends to put the color back in where it expects to see it. So next time you're outside, pay particular attention to nighttime colors. You'll be surprised at how little color you actually see.

Illustration 69

You can increase your chances of seeing very dim objects like shooting stars (meteors) by allowing your eyes to get used to the dark for fifteen or more minutes and then looking out of the side of your eye. The most sensitive part of your eye (retina) is not the central area, where most of your vision is accomplished. It's off to the side, where the rods are more highly concentrated. That's why you will see faint objects "out of the corner of your eye" better than when looking directly at them.

176

So when all else fails, try moving your head in a slow scan of the sky, aware that you may catch a fleeting glimpse of a dim star or meteor indirectly.

. . . .

TRIVIA: Despite popular belief, nothing can "see in the dark." Strictly speaking, dark means an absence of light, and with no light there is no energy reaching the rods. Even most bats rely on some amount of light to see at night. Only some species of bats use sonar for location.

. . . .

Falling Over the Horizon

Beep, beep, beep . . .
—*Sputnik I,* 1957

There are few dates in time when scientific achievements have changed the course of history. One of them is October 4, 1957. On that day the Soviet Union inaugurated what would later be called the "space race" by launching the world's first manmade satellite into orbit. Though feeble by today's standards, the faint beeping of the 184-pound steel sphere *Sputnik I* set off political shock waves around the world that still reverberate today. Catching the American public off guard, Sputnik sent President Eisenhower scurrying to close the Soviets' lead in space; a lead, we discovered later, that was as fictitious as the proverbial ether.

This technological feat, causing a manmade moon to circle the Earth every hour and a half, shocked Americans so completely that President Eisenhower had to educate everyone about the basics of "outer space." Just what was this strange place? Eisenhower instructed his science adviser, James Killian, to prepare a lesson for the American people, an "Introduction to Outer Space," for the nontechnical reader. "This is not science fiction," wrote Eisenhower from the White House. "This is a sober, realistic presentation prepared by leading scientists."

The first nontechnical explanation tackled by White House scientists was the answer to a question hardly ever thought about in those days: "Why do satellites stay up?" Dr. Killian pointed out that the basic laws governing satellites "have been known to scientists ever since Newton" but that they "still seem a little puzzling and unreal to many of us. Our children will understand them quite well."

Do they? What about the rest of us? We've all had about thirty years to find out for ourselves, so if you or your kids aren't sure, here's your opportunity to catch up. Presented for your consideration, and

178

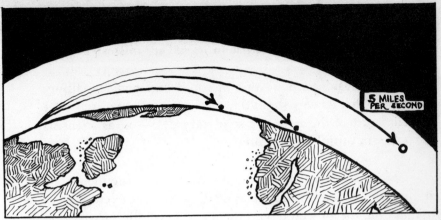

Illustration 70
Thrown at normal speed, a baseball will fall back to the earth.
But thrown fast enough, the path of the falling baseball will
match the curve of the earth and the baseball will be in orbit.

assisted by Illustration 70, is the explanation right out of the
president's report (with a few embellishments). It was wonderful then,
and it's still great today.

We all know the harder you throw a stone the farther
it will travel before falling to Earth. If you would imagine
your strength so fantastically multiplied that you could
throw a stone at a speed of fifteen thousand miles per hour,
it would . . . in fact easily cross the Atlantic Ocean before
earth's gravity pulled it down. Now imagine being able to
throw the stone just a little faster, say about eighteen
thousand miles per hour. What would happen then?

The stone would again cross the ocean, but this time it
would travel much farther than it did before. It would travel
so far that it would overshoot the earth, so to speak, and
keep falling until it was back where it started. Since in this
imaginary example there is no atmospheric resistance to
slow the stone down, it would still be traveling at the
original speed, eighteen thousand miles per hour, when it
got back to its starting point. So around the earth it goes
again.

179

In a sense you might argue that the satellite is falling over the horizon, since it is indeed falling all the time, but it is traveling so fast—five miles per second—that in its fall it "misses" the earth.

From the stone's point of view, it is continuously falling, except that its very slight downward arc exactly matches the curvature of the earth and so it stays aloft—or, as scientists say, "in orbit"—indefinitely.

Of course, since the earth has atmosphere, neither stones nor satellites can be sent "whizzing around the earth at treetop level." The friction from the air would burn them up. So satellites have to be lifted high above atmospheric friction.

It is the absence of atmospheric resistance plus speed that make the satellite possible.

Interestingly enough, weight has nothing to do with a satellite's orbit. A feather, if released from a ten-ton satellite, would follow the same path in the airless void.

There is, however, a slight vestige of atmosphere even a few hundred miles above the earth, and its resistance . . . however slight . . . has set limits on the life of all satellites launched to date [1958]. Beyond a few hundred miles the remaining trace of atmosphere fades away so rapidly that tomorrow's satellites should stay aloft thousands of years, and perhaps indefinitely.

So it's possible to fall "forever" and never hit the ground. Perhaps Killian was prophetically referring to today's communications satellites, whose orbits are over twenty-two thousand miles out in space, where no atmosphere exists to slow them down. Their useful lifetimes are limited, not because of orbital decay—they might stay in orbit for thousands of years—but because they will run out of maneuvering fuel eventually or become obsolete or worn out—just so much expensive space junk circling the earth.

Why not save them? Space Shuttle astronauts have made efforts to repair damaged satellites, but technology moves so quickly that, like yesterday's automobiles, yesterday's satellites are inferior to tomorrow's.

• •

TRIVIA: The moon's orbital speed is only two thousand miles per hour. Even though an artificial Earth satellite needs to achieve a speed of eighteen thousand miles per hour to go into low Earth orbit, the higher a satellite goes, the less speed it needs to stay in orbit once it gets there. Communication satellites, parked about twenty-three thousand miles away in space, travel at about seven thousand miles per hour, while the moon, about a quarter of a million miles away, mopes along at a speed the Concorde exceeds crossing from New York to Paris.

Down in a cave you would experience less gravity than at the earth's surface. Since you are below the earth's surface, some of the earth's mass is above your head, pulling you upward and canceling the effect of some of the mass below your feet pulling you downward. In a cave at the center of the earth there would be no gravity at all. You would be "floating," weightless. The surface of the earth is the place where gravity is the strongest. Go up or down and gravity gets weaker.

• •

Our Place in the Universe

There is no more humbling an experience than going out into the country, looking up at the clear night sky, and observing the countless numbers of stars that fill the heavens. Grandiose feelings about ourselves or our little planet are quickly lost under the vast sea of twinkling lights. The ancients thought that all the planets and stars were fixed on a dome, like the ceiling of a planetarium. Of course, we now know they are not. Even stars that seem fixed in the sky, like those in the Big Dipper, vary in closeness to the earth by many light-years. But who can fault those early astronomers? Without the aid of powerful telescopes and planet-hopping space probes, how were they to know about galaxies, black holes, cosmic dust, or rings around Jupiter and Uranus—or even about the existence of our three outermost planets, Uranus, Neptune, and Pluto—that make our night sky a much more interesting place to view? What better time than under a canopy of stars to take a tour of the night sky as we know it today?

Let's begin at home, the third planet from the sun: Earth. The earth is just one of nine planets and more than forty satellites that orbit the sun, our closest star. If the sun were a basketball, the earth would be a pea about thirty yards away. Pluto, the outermost planet, would be a speck of dust three quarters of a mile from the sun.

Raining Acid

The earth occupies a unique position in our solar system. Had it been a bit closer to the sun, its surface temperature might have been too hot and hostile to support life. Take, for example, our sister planet, Venus.

This brilliant morning or evening "star" is forever shrouded in clouds of deadly sulfuric acid. Its carbon dioxide–laden atmosphere locks in the sun's rays and, in a planetary "greenhouse effect," the

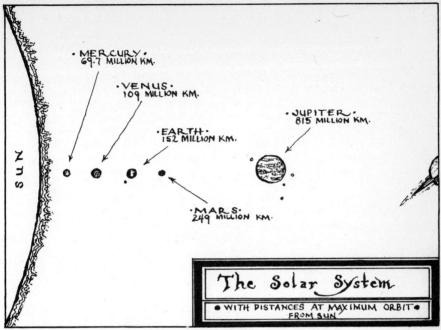

Illustration 71

sunlight is converted to scorching heat that puts the Venusian surface temperature at more than 900 degrees Fahrenheit—hot as a self-cleaning oven. Not very conducive to life as we know it.

Venus' neighbor Mercury is even closer to the sun. Mercury takes only 88 days to complete one revolution around the sun, but its quirky orbital path makes its day an extraordinarily long one: 176 Earth days in duration. Yet Mercury takes only 59 days to rotate once about her axis. What this means is that the planet makes three rotations about its axis for every two revolutions about the sun. If you were able to observe the sun from two points on Mercury's equator—longitude 0 degrees and 180 degrees—you would see the strange path it takes. The planet's weird orbit causes the sun to slow down its journey across the sky as midday approaches. Then it reverses direction for a short distance, stops, changes course again, and resumes its original track with increasing speed until it sets.

This extended "high noon" of the sun during the hottest part of the day drives the temperature over 800 degrees Fahrenheit. While nighttime temperatures can drop to below −300 degrees Fahrenheit.

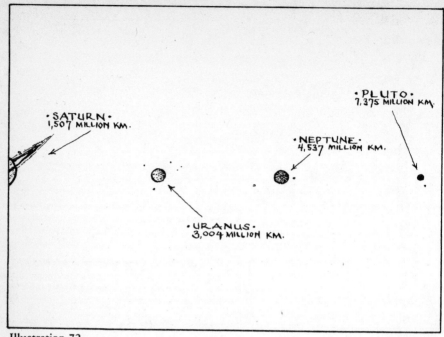

Illustration 72

Due to its proximity to the sun and lack of mitigating atmosphere, Mercury's temperature swings are the largest of any planet in the solar system—a typical night-day difference is about 1,100 degrees Fahrenheit. The tiny planet is slowly roasting like beef on a barbecue spit. With daytime temperatures above the melting point of tin, lead, and zinc (800 degrees Fahrenheit) and nighttime temperatures colder than dry ice, Mercury is hardly a place where life exists.

The fourth planet from the sun (Earth is the third) is Mars. Mars, perhaps the most famous planet and the subject of countless science fiction books and movies, is just far away enough to be too cold to support life. Its low-pressure atmosphere is thin enough to boil your blood (see explanation of boiling in Chapter IV, ". . . Cauldron Bubble"). Yet it is a planet of spectacular landscape: On one immense, dusty red plain rises a snowcapped volcano three times the size of Hawaii's Mauna Loa, with a base that would cover the state of Nebraska. Astronomers have observed that Mars has a considerable amount of water locked up in its polar caps. And despite its unbreathable atmosphere, Mars is not so terribly different from places

184

here on Earth such as Antarctica. Given the right winter clothing (including pressurized space suit) and shelter, we would be able to survive on Mars as long as food and supplies held out, a feat we could never accomplish on the impossible worlds of Mercury and Venus.

Extraordinary as these planets—Mercury, Venus, and Mars— seem, the five outer planets are even more bizarre and less hospitable. While the inner planets are composed of mostly solid material, the outer ones are made mostly of gas and liquid, with perhaps a solid central core. One theory explains this difference. During the formation of the solar system, the solar wind, emanating from the sun, blew the light material—hot gaseous matter—farther out into space, where it coalesced to form the larger gaseous planets, leaving behind the heavier matter to form the inner planets.

Jupiter, for example, is a huge ball composed mostly of hydrogen gas, so big more than a thousand Earths would fit inside it. Lacking any apparent solid ground, Jupiter has no visible surface on which a spaceship could land. One would just sink through a thick outer atmosphere topped off by clouds of ammonia. Multicolored bands in the atmosphere—Jupiter's hallmark—can be seen with the aid of a small telescope. Below the atmosphere lies a vast sea of liquid hydrogen some fourteen thousand miles deep, covering the entire planet. At the center of the planet we should find (should, because we can only theorize) a rocky core enormously hot: 86,000 degrees Fahrenheit. This fiery center may help explain why Jupiter emits twice as much heat as it absorbs from the sun. While observing this fascinating planet through a telescope or high-powered binoculars, you might notice that Jupiter appears to be flattened; it's not exactly spherical, but bulges at the equator. This is no optical illusion. The earth also bulges at the equator, due to the relatively high speed of its rotation. Jupiter rotates two and a half times faster than the earth, producing a very large equatorial bulge, about twenty-three times greater. This rapid rotation has important consequences. It means that this enormous planet, the largest in our solar system, has just a ten-hour day—less than half the twenty-four-hour spin rate of the earth. And the spinning of so tremendous a planet produces incredible planetary weather: Huge storms the size of continents may rage for centuries. Jupiter's Great Red Spot—visible through a backyard telescope—may be a Jovian tropical "hurricane" that's been raging for over a hundred years but appears now to be shrinking.

Jupiter is surrounded by sixteen moons and a newly discovered ring—unnoticed until photographed by the space probes *Voyager 1* and *Voyager 2* in 1979. Jupiter's moons are some of the most intriguing heavenly bodies to be found anywhere. Io is the most volcanically active body—planet or satellite—in our solar system. Up to eight volcanic plumes have been seen there at one time. The moon Europa is covered by a maze of thin lines that look a lot like the fabled canals of Mars. But the surface of Europa is so smooth it has been compared to a badly scratched crystal ball. The moon Ganymede—largest in our solar system and larger even than Mercury—is crisscrossed by an enormous system of "highways," bundles of grooves that have yet to be explained. Callisto, Jupiter's outermost moon, is completely pockmarked with craters, the most densely cratered body known in the solar system. The moons of Jupiter are so unusual and varied that astronomers tend to think of the giant planet as the center of a minisolar system of its own.

The rings of Saturn have always caught the attention of astronomers, but their picture of the planet was forever changed when, in November 1980, images sent back by *Voyager 1* showed these rings to be composed of thousands of enormous, thin ringlets spaced closely together like grooves on a phonograph record.

In addition to the other planets—Uranus, Neptune, and Pluto—our solar system is home to thousands of small planets called asteroids that fill the large gap between Mars and Jupiter. Astronomers believe the asteroids may be the elements of a planet that was to be but never formed. According to prevailing theory, great gravitational forces of giant Jupiter, pulling on the asteroidal chunks, prevented the material from ever solidifying into a single, cohesive planet. Comets (from outside the solar system) come and go in their elliptical journeys, leaving behind dusty tails that fall to the earth as meteorite showers visible to the naked eye.

Looking Back in Time

Out beyond our solar system is the vast expanse of deep space. Our nearest neighbor—the star system called Alpha Centauri—is

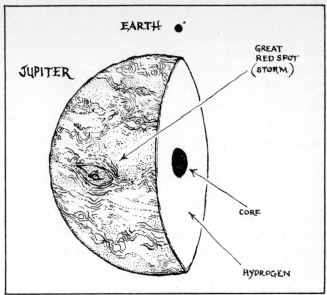

EARTH

JUPITER

GREAT
RED SPOT
(STORM)

CORE

HYDROGEN

Illustration 73

beyond our ability to visit. Even a light beam traveling at 670 million miles per hour takes more than four years to get there from Earth. Thus we say that Alpha Centauri is a little over four light-years away. Other visible stars are much farther away. The brightest one, Sirius, is nine light-years distant; the second brightest, Canopus, is over five hundred light-years away. What this means is that when we look at starlight we are actually looking back in time. The light from Alpha Centauri that we see today left that star system four years ago. The light from Canopus left more than five hundred years ago. We are seeing events on Canopus that occurred before we were born. And by looking back deeply enough into space, we can see objects and events that are billions of years old—older than the earth, older than our sun and solar system. That's why astronomers are constantly making more and more powerful telescopes—so they can see deeper into space and farther back in time to learn more about the origins of our universe.

In the process, they have discovered just where our tiny planet sits in our galaxy. All the stars visible to the naked eye and hundreds of billions of others make up our home galaxy, called the Milky Way. The Milky Way Galaxy is spiral-shaped, and best estimates place one hundred billion to four hundred billion stars within it. The galactic spiral is about one hundred thousand light-years across and may be surrounded by a much larger sea of dark matter whose character astronomers don't quite understand. Dust and gases take up space between the stars. It is somewhat pretentious to think that we live on the only planet in the galaxy, and many astronomical theories predict the existence of other planets.

But new planets are hard to find for many reasons. They don't give off much visible light of their own, they just reflect the light shining on them from the "sun" in their solar system. They're so far away it's impossible for the naked eye to see one. Theoretically, a very powerful telescope might be able to pick out the reflected, visible light of a distant planet but none has been able to.

A more promising technique is to use a special telescope to look for invisible radiation given off by the planet itself. All objects, no matter what their size or temperature, give off a certain amount of heat in the form of infrared radiation. (An ice cube, though freezing cold to the touch, is still "warm" in the sense that its temperature is above absolute zero. So it gives off heat in the infrared range.) Planets also take some of the light shining on them and reradiate it to space as infrared radiation. Infrared is not visible to the naked eye but can be detected by an infrared telescope. By pointing such a telescope at a star and looking for the infrared signature of the star, astronomers believe that they can find a "cold" planet shining in the infrared wavelength in proximity to the star. The idea is to look at the spectrum of radiation given off by a well-studied star and see if there is an extra bit of radiation that can't be explained by the presence of just the star and might indicate it has a companion: a planet.

In fact, using an infrared telescope, astronomers think they have located a few good cosmic candidates, among them the stars Beta Pictoris and Giclas 29-38. There appears to be an excess of infrared coming from Beta Pictoris—too much to be accounted for by the star itself—and the infrared signature shows that Beta Pictoris may be surrounded by an asymmetrical disk of dust. The temperature of the disk is much less than that of the star and the fact that the disk is not exactly round suggests that the asymmetry might be due to an unknown, substellar object moving around the star in a noncircular orbit.

A Brown Dwarf?

There is stronger evidence of a discreet object orbiting an aging star Giclas 29-38 located just fifty light years away (a bit less than three hundred trillion miles) in our own backyard, the Milky Way galaxy. However, the temperature and size of the object makes its identification a problem. It is too warm to be what we consider a planet but not massive enough to be a star. It is hotter and more massive than any of the known planets in our solar system, making comparison with even the largest, Jupiter, problematic. The object appears to be about 50 percent larger (in terms of mass) than Jupiter and ten times hotter. The fact that this object is giving off so much heat—on the order of about 1,700 degrees Fahrenheit—indicates that it might be something never observed before: a brown dwarf, a celestial body that is neither star nor planet but something in between.

For many years, astronomers have speculated about the existence of brown dwarfs, but the discovery of one has never been confirmed. One brown dwarf seemingly disappeared after its initial sighting and has never been seen again, casting doubt on its existence. This time, the evidence is concrete and the object meets all the criteria the theories dictate for mass and heat necessary for a brown dwarf.

The discovery would be a tremendous boost to the understanding of the universe we live in. First, it would help astrophysicists calculate the mass of the universe. Since they believe that perhaps as little as 10 percent of the mass of the universe is visible, part of that dark mass might be in the form of hard-to-see brown dwarfs.

Second, the size of the object begins to tell us how small an object can be (in terms of mass) before it develops into a star. Jupiter, with its "planetary" system of moons, appears to be a star that almost was: bigger than a conventional planet but too small to have become a star. An ongoing thermonuclear fusion reaction is at the heart of a star. This brown dwarf candidate, 50 percent more massive than Jupiter, is still not massive enough for the compression and heat caused by its own self-gravity to set off thermonuclear fusion at its core. So an object, to have its nuclear fires lit and become a star, must be larger than this one.

Where does our solar system reside? Not in any place of distinction. Our sun sits off to the side, about thirty thousand light-years from the center of the Milky Way; not on any of her majestic spiraling arms but off on a "side street" of stars. Thus our Spaceship Earth is just an unobtrusive speck of rock, metal, and glass circling an ordinary star unceremoniously located in the backwater of a vast mass of stars just like it.

Beyond the Milky Way

Beyond our Milky Way Galaxy lie large, empty voids called intergalactic space. Free of most matter, these voids might contain a single atom per cubic yard of space. Closer to home, tagging alongside our galaxy, are two small galaxies called the Magellanic Clouds. Visible in the Southern Hemisphere, the Magellanic Clouds give astronomers a chance to study a relatively close galaxy (150 to two hundred thousand light-years away) that has evolved independently of our own.

If you go out on a very dark and clear night, you might be able to see our nearest large neighbor, a galaxy in the constellation Andromeda and called by that name. A little more than two million light-years away, the Andromeda Galaxy is a magnificent spiral that looks very similar to our own. Its beauty and easy visibility make it a prime viewing attraction for amateur stargazers.

Beyond Andromeda and far beyond the lenses of backyard astronomers lie the oldest and most mysterious objects in the sky: quasars. Quasars look like stars but behave differently. A quasar—a contraction of the original name "quasi-stellar radio source"—appears to be in the throes of a violent reaction. Powerful "jets" of high-speed material are being spewed out by an unseen mechanism at its center. Quasars not only shine with the brightness of a hundred galaxies but also send out very strong invisible signals that can be picked up only by radio telescopes. Quasars are the most distant and energetic objects known. Observing them is like looking back in time about thirteen billion years, and seeing the universe as it was about a few billion years after a colossal explosion that formed all matter and energy—the Big Bang.

Quasars and galaxies surround us in all directions. Galaxies appear to take several shapes but astronomers have classified them into three major ones: elliptical, spiral, and irregular. Elliptical galaxies

190

show no structure other than an extremely smooth and even distribution of light. They come in a variety of shapes that differ in the amount of flattening. They may be spherical or pumpkin shaped, or almost as flat as a disc. Some elliptical-shaped galaxies are surrounded by hot X-ray halos.

Spiral galaxies are recognized by their central bulge outside of which are spiral arms. Our Milky Way is a spiral galaxy. Some spiral galaxies, though appearing to be filled with bright stars, contain much greater amounts of "dark matter," which we can't see, than the glowing stars in them that give us such pretty pictures. The dark matter may extend many times as far as the visible stuff.

Irregular galaxies are odd shaped, exhibiting no particular structure. The Magellanic Clouds belong to this type.

Supercluster Complex

Galaxies may form groups of filamentary and sheetlike clusters that move together through the expanding universe. One astronomer has found evidence that our own Milky Way galaxy is part of a flat, oblong "supercluster complex" that encompasses millions of galaxies and stretches one tenth the distance across the observable universe. This would be the largest known structure in the universe.

It takes one billion light-years—about 10 percent of the age of the universe—for light to cross the length of the complex, suggesting that its pattern must have been laid down shortly after the birth of the universe, according to its discoverer, Brent Tully of the University of Hawaii's Institute for Astronomy, in Honolulu.

According to Tully, the evidence for the supercluster complex, one hundred times more massive than any previously known structure, would suggest that, on this immense scale, galaxies are not randomly distributed throughout the cosmos but are instead clustered in space in a way that is not anticipated by current conventional theories of galaxy formation. The conventional theories say that the universe should have a uniform distribution of matter as a result of the Big Bang, the theoretical explosion that created all matter and energy ten billion to twenty billion years ago. As in a small explosion, matter from the colossal Big Bang should be spewed out uniformly in all directions. But the discovery of the "lumping" of galaxies into un-smooth clusters has forced astrophysicists into questioning what unknown events following the Big Bang could have caused the formation of clusters and superclusters. As more superclusters are discovered—at least four others have reportedly been sighted—astrophysicists will get a better understanding of the structure of the universe and our place in it.

• •

TRIVIA: About 50 to 90 percent of the mass of the universe is dark. Bright stars and glowing gases make up only a fraction of what is out there. The rest of the matter that is present is in a form invisible to our telescopes. It may be composed of exotic particles yet to be discovered, it may be "cold" stars or miniblack holes, but most of it is clumped around galaxies. Early in the history of the universe this "dark matter" may have played a key role in directing the formation of the galaxies and in influencing the present structure of the cosmos.

Statistically speaking, life in outer space is almost a certainty. The number of stars estimated to exist in the visible universe is equal to 10 followed by twenty-one zeros (10,000,000,000,000,000,000,000). Among those stars it is statistically likely that some will have planets, some of those planets will have enough light and heat, and some of those will be situated just the right distance from their star to support life. If each of those three possibilities is rated a one-in-a-million chance of being true, there should be about ten thousand planets inhabited by living beings in our universe.

• •

You Can't Climb Out of a Black Hole

I think that there should be
a law of Nature to prevent the
star from behaving in this absurd way.
—Sir Arthur Eddington

If it weren't for astrophysics, our vocabulary would be missing one helluva great phrase: black hole. Did something disappear? It fell into a black hole. Where are the car keys? A black hole must have sucked them up. Black hole says it all: a strange, unfathomable object that drifts by, gobbling up anything near it. Isn't that what happens in space?

This description is not too far from the truth. A black hole (a name coined by physicist John Wheeler) is an object in outer space with a gravity so strong that nothing can escape from it, not even light.

A black hole is a star that has collapsed, condensed, and become invisible. How has this happened? Gravity is the key.

The gravitational force of an object depends on its mass. In addition, the closer you get to the center of the mass of the object, be it a planet, a star, or a basketball, the more powerful the gravity becomes.

As an illustration, imagine that you are a person weighing two hundred pounds and standing on the earth. You would be (approximately) four thousand miles from the center of the earth, which is the source of gravity. Suppose you could build a ladder tall enough so that you could climb four thousand miles high. At the top of the ladder, you would be twice as far from the center of the earth (eight thousand miles), so you would weigh much less—only one fourth what you did on the earth's surface—or fifty pounds.

Come down from the ladder and look at it another way. Let's say that instead of doubling your distance from the center of the earth by climbing a ladder, we could double your distance by enlarging the earth to twice its size. To do this, we double the distance between all the atoms in the earth. No new matter has been added. The original matter has been merely spread out making the size of our planet double.

What has happened? The surface of the earth is now eight thousand miles from the center. And you, standing on a bathroom scale on the surface, weigh fifty pounds.

This conduct of gravity also works in the opposite direction. If we could squeeze the Earth to half its original size, you would weigh four times as much—800 pounds—at its surface. Squeeze it even more, to one tenth its initial size, and standing on its surface you would weigh ten tons!

It's becoming evident that if an object could be compressed small enough, it would be possible to produce very powerful gravitational forces. And if that object was very massive to begin with, the resulting gravitational force would be almost unimaginable.

Consider a situation in which you had a very massive object (like our sun) and then cosmically squeezed the object until its size was no larger than a couple of miles in diameter. What would happen? All of the star's mass would still be there, so the star still would have a gravitational pull. But since the mass is now concentrated closer to the center of the object, the force of that gravity is much stronger—over fifty billion times stronger! Now, if our object was massive enough, nothing could escape from its gravitational pull: not a rocket ship, not even a light beam. We would have a black hole—black because we could not see it, a hole because it would suck in matter like a swirling gravitational drain.

A black hole is hard to picture and difficult to contemplate. This is because according to the theory of relativity, a very strong gravitational force of the magnitude found in a black hole influences space and space-time. Einstein's general theory of relativity says that gravity makes space curve. The path of an object that would normally be straight in open space is made to curve as the object approaches a planet or star—anything with a strong enough gravity. In the case of a black hole, which has a superstrong gravitational pull, space is not just mildly curved but bent right into the center of the black hole.

194

Any object that comes near will be drawn into the funnel and will "roll" down the funnel toward the black hole at the bottom (Illustration 74).

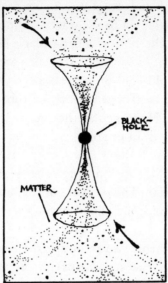

Illustration 74
A black hole warps the space around it. Objects coming close are sucked down the hole as easily as water rolls down a funnel.

The effect of gravity on time is also dramatic. Place a clock in a weak gravitational field, like that of the earth, and it is very difficult (but possible) to detect the effect of gravity on time. But place the clock in the immense gravitational field of a black hole and the effect on time can be readily seen, depending upon the location of the observer. To a space voyager falling into a black hole, the powerful gravity would greatly speed up his perception of things. According to his observations, he would be quickly sucked to the "bottom" in great and final haste. But for those of us watching his fateful journey from outside the influence of the black hole, the event would appear to be just the opposite. Our unfortunate astronaut would seem to be moving very slowly. To us it would seem that his fall was becoming slower and slower the closer he came to the bottom of the hole. To explain why, we would say that the intense gravity of the black hole was slowing down time—as predicted by the theory of relativity—so that he would take forever to disappear out of sight. At the bottom—where all mass and energy are concentrated into an infinitesimal point—space vanishes and time ends. The laws of physics as they apply to the outside world are suspended and we have no way of knowing, beyond theory, what is occurring at the bottom of the black hole.

How can a bright, shiny star turn into a dark, forbidding black hole? The answer is old age. Stars are mostly made of hydrogen gas, the same lightweight stuff that used to keep blimps or dirigibles like the *Hindenburg* aloft. Hydrogen is the fuel that keeps a star shining. Deep inside each star is an ongoing thermonuclear reaction, sort of a continuous H-bomb explosion that turns this gas into energy: heat and light. As long as the star is "burning" the hydrogen, it's involved in a continuous tug-of-war. The pressure of the heat inside forces the star to expand, to blow up like a balloon into its starlike, round shape. But at the same time, the star's own gravitational pull is trying to make the star shrink from within. So as long as the heat is on, the tug-of-war is at a stalemate and the star keeps its proper shape. Without its nuclear fires, however, the star would give way to gravity and collapse like a balloon losing its air.

But as a star gets older, it begins to cool off. With the loss of heat, this aging giant cannot produce enough internal pressure to counteract the shrinking forces of gravity, so it begins to collapse and shrink in size. But while the star may be shrinking, it isn't losing any of its mass; the hydrogen is still there, it's just being supercompressed. That means that all of the mass of the star is coming much closer to its center, in effect concentrating the gravity into one small location. Small stars will shrink into what is called a "white dwarf," an object about the size of the earth but whose nuclear fires have gone out. Larger stars self-destruct in a blaze of glory called a supernova, blowing away much of their matter in the process.

But if what's left is massive enough —about 1.4 times the mass of our sun—the remaining matter may be on its way to becoming a black hole. Let's say our star, as in Illustration 70, gets so small that its billions of billions of billions of tons of matter were compacted into an object just one mile in diameter. The gravity would be so strong on its surface that nothing could ever escape it, not even its own light.

The object would be there, but you wouldn't be able to see it. Objects close by would be sucked in and disappear down the "black hole."

Astronomers studying spiral galaxy NGC 4151 in the constellation known as the Hunting Dogs believe that at the core of this rotating spiral there is an extremely massive black hole fifty to a hundred million times the mass of our sun. Black holes are very mysterious, and because they cannot actually be seen, it is very difficult to know if we are actually looking at one or if they really exist. This makes the claim for discovering a black hole very controversial. But it's believed that black holes may lie at the centers of galaxies, perhaps even in our own Milky Way. Out at the edges of our universe lie fantastically shining quasars. Astronomers believe, for lack of a better explanation, that the immense power needed to drive these quasars may be coming from black holes at their centers. How do you know if you've found a black hole? Any time an unseen object that is many times more massive than the sun is discovered, it is suspected of being a black hole. While

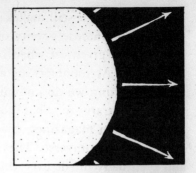

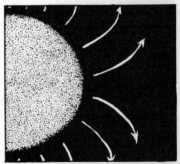

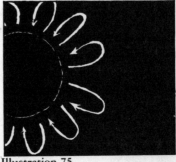

As a massive star dies and collapses to become a black hole, nothing can escape its immense gravity. Not even its own light can depart; it merely "falls" back down.

Illustration 75

it can't be seen, it still makes its presence known, or suspected, by the effects of its intense gravity on other matter near it. For example, X rays may be given off when intensely heated matter is sucked into a black hole. Astronomers are able to detect this radiation.

One thing we do know is that black holes are nothing to worry about. Contrary to what you may see at the movies or read in science-fiction magazines, black holes are not the doomsday vacuum cleaners of the universe. They are not going to float around randomly and eventually inhale everything: stars, galaxies, and planets, including ours. In fact, just a short distance away from a black hole the gravity is quite tame. If the earth were squeezed enough to make it the size of a golf ball, it would become a black hole. But the total amount of gravitational pull of the "earth black hole" would still be the same as that of present-day earth. To be sucked into the earth black hole you would need to come within an inch of the "golf ball." There the gravity would be so strong, the space so warped that escaping would be impossible. But park your spaceship four thousand miles above the golf ball—equal to the distance from the center of the earth as we know it to the surface—and you would feel the same gentle gravitational tug felt normally on earth.

• •

TRIVIA: Stars can be shrunk into black holes because matter is mostly empty space. The distances separating the electrons from the nucleus of an atom and from each other are greater in scale than the distance separating the planets of our solar system. Squeeze out all the empty space in the atoms of the Empire State Building and the 102-story structure would shrink to the size of a toothpick. But it would still weigh the same.

• •

Warp Speed, Scotty

Next time you feel like relaxing and someone harps on you to get up and get moving, tell him or her you're already moving fast enough.

The sun transversing the sky is a reminder that we're all passengers on a ship moving at incredible speed. This is easy to forget because the ride is so smooth we hardly know we're in transit, and since there are no trees whizzing by our planetary window, no references exist to gauge our motion. But in truth, not only are we spinning like a top but our planet also is whizzing through space at a breakneck pace, as shown in Illustration 76.

Our planet makes one complete rotation on its axis in twenty-four hours. So each of us, residing on the planet and under the influence of gravity, is moving along with it. At the equator that speed is 1,040 miles per hour. (Why are people living in San Francisco and São Paulo moving slower? See in Chapter V, "Tornado in the Drain.")

But in addition to spinning on its axis, the earth is also revolving around the sun, quite rapidly. A quick computation shows how fast. At a distance of about ninety-three million miles from the sun, the earth travels along an orbit almost six hundred million miles long. Going that distance in one year—365 days—requires an average speed of sixty-six thousand miles per hour. That's faster than anything on the earth, 3½ times faster than the Space Shuttle, and over a hundred times faster than an airliner.

Of course, the earth is linked to the sun by gravity, and where the sun goes, so goes our planet. The sun is just one of hundreds of billions of stars in the spiral-shaped Milky Way Galaxy. Each of these stars is moving. But how can this movement be described? To what reference point can we compare it? Again, we have no trees or telephone poles whizzing by, nothing we might say is standing still. So astronomers arbitrarily define a standard of rest in our section of the galaxy. They compare the motion of all the stars and arrive at an average velocity. The velocity of our solar system can now be computed, and it turns out that we—the sun and the planets—are all hurtling through space at about forty-three thousand miles per hour, headed toward the fifth brightest star in the sky, Vega, in the constellation Lyra. (Don't make any plans for dinner there—Vega is twenty-six light-years away.) But there's more to the story. Remember that our galaxy itself is spinning like a pinwheel and we're caught in that motion, located about three fifths of the distance from its center. Stars that far away take about two hundred million years to make one complete trip around the middle. Our sun overcomes that megadistance by traveling at a galactic speed: six hundred thousand miles per hour. And we, sitting at our little planet tethered gravitationally to the sun, whiz through space at unimaginable speed, oblivious to the motion that makes everything else, by comparison, seem to be standing still.

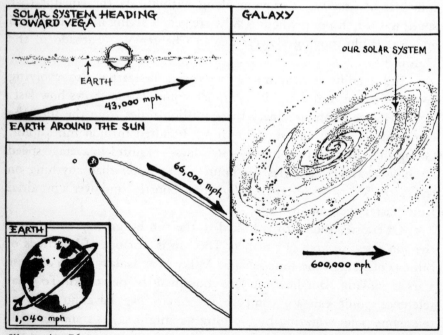

Illustration 76

VIII. The Techno-Zone

Any sufficiently advanced technology
is indistinguishable from magic.
—Arthur C. Clark

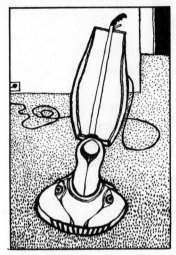
Illustration 77

There's an old *Twilight Zone* episode in which a man comes home to find that his electric appliances have turned against him like animals against the zookeeper: The electric razor comes slinking down the steps like a python looking for a good meal. The self-starting vacuum cleaner creeps across the carpet in solitary frontal assault. It's an eerie episode that strikes home because Rod Serling exploited a fear that many of us share: new technology. The irony of this fear is that not a day goes by when we don't talk on the telephone, watch television, or take a ride in an airplane; yet how many of us trust the very technology we depend on? We regard technological objects as mysterious, frightening creatures with minds of their own, eagerly waiting for the chance to break down just at the wrong moment.

But don't be intimidated. Hi-technology is more bark than bite. A little bit of knowledge goes a long way in getting the upper hand on the inhabitants of the hi-tech jungle. As Mr. Serling might have said, "Next stop: the Techno-Zone."

Illustration 78

Two Wires to the World

If there is one thing that Ma Bell used to do right in the days before the phone business was deregulated, it was making quality, lifelong telephones, telephones made to last for decades. My mother just gave back her rented phone after a good thirty years of use, and I had to practically pry the receiver from her fingers. Lots of even older phones are still being dialed by grandmas who refuse to give up a good thing. Today's phone system is different, however. You walk into any department or phone store, plunk down a few bucks, and walk out with a phone that is all yours. It's like buying a pocket radio. Some are so cheap you don't fix 'em, you throw 'em away.

In the old phone-rental days, a telephone was made to last forever because if the phone went on the fritz, a repairman had to come out and fix it at no charge. So it was in the best interest of the company to make long-lasting equipment or else it had to foot the bill.

These days if the phone person comes to your home to install a postderegulated model, you're charged a fee for the installation of a single phone jack in the basement (the jack is now called a "network interface"). In older homes or apartments where there is no interface, the first junction box inside the home is designated by the phone company as the interface. This is where the phone company's own wiring ends and yours begins. So if you want additional wiring, there's an extra charge to bring the wires upstairs and install jacks in your kitchen or bedroom. Or if you already have a phone and you want extensions in other rooms, that'll cost extra, too. Phone installation charges pile up rapidly.

But there's no need to pay anyone to do the work you can do yourself. Why? Because once you know how simple it is to hook up a phone, you're going to be shocked (not electrically, of course). If you know which end of a screwdriver is up, you can install extension phones as easily and safely as a professional installer can.

204

Since the phone works on just two wires, installation is easy. Run this pair to another room, hook the two to a wall jack, snap in the phone, and you're done! All the parts you need can be bought at a hardware store or supermarket.

First: Update Your Phone Jack

If you have an older phone, it will be attached to either a permanently wired terminal block or a four-prong plug mounted on the baseboard or walls. As long as you're adding new phones and phone cables, you might as well do a complete job and replace the old terminal blocks with the newer-style modular jacks. It's a simple matter to buy a replacement modular adapter at the store and do it yourself.

Remove the cover on the old block (the single screw in the center holds it shut). Inside you will see four color-coded wires—red, green, black, and yellow—attached to four screws. Loosen the screws a bit (don't unhook the wires) and connect the colored wires from the new modular jack to the screws on the old terminal (some modular jacks come with caps that easily snap over the screws; just read the instructions on the package). Check that your color coding is correct, and screw the cover on.

At this point it's easy to hook up two phones in one room by adding a duplex plug to the single jack and plugging each phone cord into it. But if remote extension phones are what you desire, then it's best to install new cabling between rooms. Measure the length of new cable you will need by carefully planning the route the cable will take around doors, through walls, or along baseboards. Do not run the cable under carpeting or let it dangle loosely where people will trip over it or where it will get wet. Go to a local store or Radio Shack and buy the amount of cable you need. Tacking the cable to the wall or baseboard is simple using insulated U-hooks hammered down every few feet. Once the wire is run, you're ready for the easy part: hooking up the phone.

Hook a special wiring junction box to the network interface. Buy one that has a short cable with a modular plug that can be plugged into the new interface (see Illustration 79).

Open the junction box; inside are four sets of color-coded

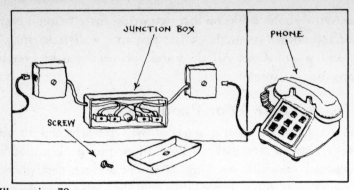

Illustration 79

A junction box helps make installing extra phones easy.

terminals, designed to be matched to the special four-conductor phone cable. As I mentioned earlier, only two of the wires—green and red—are actually needed to work the phone. One carries the dial tone and the other rings the phone. But as long as the box is open and you're working on it, it's best to hook up the other two wires—black and yellow—in case you want to add a second phone line later. The black and yellow carry the second line. They go unused unless you have a second telephone line. (One phone number means you have but one line.)

Take the new phone cable and with a sharp knife, razor blade, or wire stripper (invest in one, it's worth it) cut away the plastic cable coating. Be careful not to cut the wires inside. If you do, just cut off and discard a short piece of cable and start over again.

Find the green and red wires in the cable. Strip away about half an inch of insulation from each one. Unscrew the little screws holding down the red and green wires inside the junction box. Don't remove them, just loosen them enough to get another wire inside. Then wrap the stripped copper wires around each of the four color-coded terminals. Make sure the wire is wrapped clockwise, because that's how the screws are tightened and that makes the wire hold better. Be sure the wires don't come off the screws as they're tightened, and make sure the bare wire connections don't touch each other. Put back the cover, screw it tight, have a diet cola, and smile. Half the job is done.

At the other end of the cable, it's the same deal. Unscrew the cover from the new phone terminal and screw the terminal to the

206

baseboard or wall. Strip the insulation from the wires as you did on the terminal block and connect the wires to the screws on the plastic terminals (do not tighten). Each screw will be labeled G, R, Y, B, according to the color wire. Connect the colored wires from the jack to the marked screws. Tighten the screws, replace the cover, and you're done. Snap in the phone cord, pick up the receiver, and listen to the sweet sound of success. If you hear a dial tone but cannot dial out, it means the red and green wires are reversed. Simply switch them around. Sometimes the phone company has reversed them on the telephone pole outside.

Wall phones are a bit more tricky but nothing to be afraid of. You'll need a special wall phone jack, though. Just follow the directions on the box. Now that you know about the wiring color coding, there's nothing to hold you back.

Installing a Second Phone

If yours is a two-phone-number house—perhaps one for business and one for personal use—it's just as easy to hook both lines to a two-line telephone. Remember those spare yellow and black wires in the cable? Use them to carry the second phone line. Chances are both phone lines enter the house at the same place. Hook up a second terminal box for the second phone line to the network interface of phone line 2. It's quite easy to branch off a set of wires from the yellow and black terminals in terminal box 1 to the red and green screws in the terminal box of phone line 2. Plug in your two-line phone on the other end and you're all set. Again, check the dial tone. If it's there but you can't dial out, reverse the black and yellow wires.

Just two words of caution: If the phone should ring while you are holding the *bare wires,* you might get a *mild* electrical shock. Work holding only one wire at a time and don't hold on to any metal pipes or radiators while working.

Also, be warned: Once your friends and relatives learn of your phone installation talents, you're in trouble. The demand for your services can keep you very busy. Make sure they at least make you dinner.

The TV Gets Brains

Television is like the family car. Newer models come out every year. They're fancier and more highly engineered than last year's. But when you come right down to it, the basics of TV technology, like the fundamentals of the internal-combustion engine, have changed very little since Great-grandpa took a test ride. True, TV has been refined, but the underlying principle remains the same. That old seven-inch black and white you bought in the 1940s will still pull in the same channels it could back then (of course, channel 1, a fixture of my old RCA, is no longer in use).

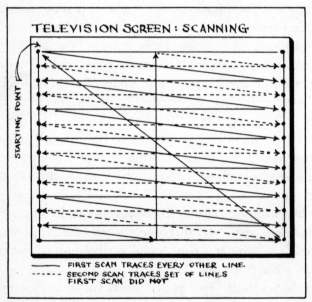

Illustration 80

Inside your picture tube, an electron beam scans back and forth from top to bottom, creating little lines. If you look closely, you can see those lines (see Illustration 80). The lines are made up of dots that glow when struck by the beam. Black and white tubes produce varying shades of gray. Color TVs mix and match three colors—red, blue, and green—to produce the rainbow (peacock?) of colors on the tube.

That's the basic idea. By playing around with the arrangement of dots, the design of the tube, or the number of lines on the set, one can alter the quality of the picture. This is what makes TV manufacturers different from one another—how they alter and improve the basic design. Some will try to make the dots as sharp as possible by inserting a "mask" through which the beams shoot. Others will fire the three color beams from one source (gun) instead of three to focus the picture more accurately.

In the end, the picture you like best is a matter of taste. No one can tell you which set looks better than the others. The differences may be so slight you might have trouble seeing them.

That's when it's time to look at the added features that folks in the days of Howdy Doody never dreamed possible.

Understanding the "Specs"

There you are, amid a hundred TV's, all playing *Wheel of Fortune* at the same time. Shoppers are jockeying for the best view. Salesmen are rubbing their hands as you approach, dollar signs in their eyes. Then like Scotty to Captain Kirk come those salesmen hi-tech questions:

"Do you want MTS or RGB? A plain receiver or receiver-monitor? Stereo, hi-fi, or stereo hi-fi? What size screen? Console or table model?"

It can be as mind-boggling as a walk through Wonderland. But you're ready because you've taken two key steps in advance: (1) you've studied the "specs" so you know what you're talking about; (2) you know exactly what you want.

"How many lines resolution?" you snap back. "Is it computer-compatible?" The salesman is knocked for a loop. He's not ready for an informed consumer. He (or she) goes looking for the manager. You've got them on the run.

The "specs," short for specifications, are the details. When you buy a car, you ask for the specs: how many miles per gallon, what's the size of the engine, the top speed. You compare one car's specs with another. It's no different with television. Only the jargon is different.

Topping the specs list is horizontal resolution, the measure of picture sharpness and clarity. Horizontal resolution is the number of vertical lines the set can draw across the face of the screen. It refers to the number of picture elements (pixels) there are in a single line of the TV signal. Don't worry if you don't understand this. All you need to remember is the higher the number, the sharper the picture.

A regular TV set will have between 250 and 275 lines. Monitor-receivers (more about them later) will have up to four hundred lines or more. Look for those higher numbers for a better picture.

But even though your TV may be able to produce more than four hundred lines of resolution, TV stations can broadcast no more than 336 lines. Does this make the spec number meaningless? No. Still use it as a comparison of TV quality. Because as TV signals get better, your TV set won't become obsolete.

In addition, if you use your set for more than just receiving TV signals, you will want higher resolution. Videodisc players put out up to four hundred lines. Videotape recorders can or will top this number. So if you're future-minded, invest in higher resolution. But don't go hog wild and invest in a set that puts out more than 450 lines; you'll probably never need it unless you use your set as a computer monitor.

Now you're interested in looking for another spec: RGB.

RGB: The Ultimate in Color

Now that you're deeply into color, it shouldn't take much guessing to figure out that RGB stands for red, green, and blue—the three primary colors in television. If a television set has an RGB input, that means it can double as a computer monitor. Many of the internal electronic guts of the TV receiver are bypassed so that the computer signal goes straight to the three color guns in the picture tube. So the resolution of the monitor is limited only by the TV itself and the output of the computer. The computer-generated picture should be much better than that of a simple TV station.

Many inexpensive computers do not have RGB outputs. They have a composite video output, which in effect takes the computer signal and makes it into a TV signal that cannot be fed into the RGB input of your TV. You have gained nothing. To take advantage of your monitor's RGB resolution, make sure your computer has an RGB output as well.

Cable-Ready

The idea of cable television comes from the early days of TV when people who lived outside big communities had trouble receiving signals broadcast from far away. They banded together, hooked up a big antenna atop a high tower, and ran cables to all their sets. This quaint "community antenna television" (CATV) evolved into the multibillion-dollar cable TV industry that calls you on the phone to ask for your subscription.

Most of the new televisions made today are "cable-ready." All this means is that your TV is equipped to tune into the extra stations supplied by the cable. All you need to do is plug the cable into the back of the set. How many channels you can get depends on the TV. Some can receive 68, others 82, the heavyweights can pull in 178. Whether there is anything worthwhile to watch on all those channels is another story. But there they are if someone wants to program them.

The irony of all this is that if you subscribe to cable, they will supply you with a cable box—a tuner that you plug into your television. Since you use the box to tune the stations, the cable tuner on your television goes unused. However, if you run extra cable to other sets in the house, you can use the cable tuners on those sets. But since the signals on the premium cable channels may be scrambled (HBO, Showtime, etc.), you can only see those scrambled stations on the TV with the box attached.

Stereo TV

It's been called the biggest development since color television: stereo television or MTS (multichannel television sound). The name means just what it says. Along with your picture comes great sound, right out of your stereo. Actually, TV sound has always been of excellent quality. It is really a high-quality FM radio signal. But the speakers on TV sets are so poor that good sound is wasted on them. That doesn't matter, since TV producers never pay much attention to the sound quality they record. If the sound system were better, would TV sound improve? It remains to be seen now as stereo TV comes of age.

But before you buy a stereo TV, make sure you need one. Make sure a TV station in your area is broadcasting in stereo. If you'd rather save money and want to wait, your best bet is to purchase a "stereo-ready" television that will accept the decoder. Also, make sure you buy a TV with an MPX jack. This makes the addition of the MTS decoder very easy later if you want one.

Refitting an old set without the MPX is more difficult and will require special parts. If you're in the market for a new TV and someday think you might want to listen to stereo, go with a set that at least is "stereo-ready."

On the bright side, many VCRs are fully MTS-equipped. That means you don't need to buy a separate stereo TV. You can use your VCR to do the job. But don't be confused. A VCR *must* say that it is *MTS-equipped.* Many VCRs advertised as "stereo" record stereo "music" but not stereo (MTS) television. *Caveat emptor.*

Supersets: Monitor-Receivers

If money is no object, then you probably will want to go for the best sound there is. Now we're talking monitor-receivers. These top-of-the-line TVs offer just about all the features: RGB, MTS, remote control, and cable-ready. Their resolution is usually 330 to 450 lines and they have inputs for a computer, stereo system, one or two VCRs, a laser disc, and camera. The number of features depends on what you want to pay, which will be plenty.

If you want the ultimate, there is only one more step up:

Smart TVs

It had to happen. Sooner or later someone was going to plunk a computer chip into a television set and create a smart TV. When you consider that a computer is really nothing more than a computer chip

Illustration 81

with a monitor attached, a TV with a chip attached is virtually the same thing: the ultimate smart television.

Many smart televisions come as monitor-receivers with added features. In addition to the conventional remote control and channel scanning, smart TVs open up a whole world of communication that goes far beyond just watching programs. You can remotely control lights and appliances. Merely push the right buttons and your set will make sure your lights are off in the living room or your coffee pot is started for breakfast. Away from home? Forget to turn on the lights or the burglar system? No problem. You can call in the commands via any Touch-tone phone. Want to leave a message for Dad or the kids? Just key your note into the set and they'll read the message on the screen when they get home. Need an appointment calendar? Your TV "memo pad" is at your service. Up to eighty pages of important dates and appointments plus a ninety-nine-year calendar are at your command. Want to prevent the kids from watching X-rated movies? Lock out the reception of certain channels.

Digital Delights

The computerization of television would hardly be worth the effort if the computer could not be harnessed to improve the picture. That's where digital television comes in. By converting the incoming TV signals into the ones and zeros of computer language, the picture can be controlled by the computer.

This means you can do many of the special effects seen on TV commercials and sporting events. For openers, let's say you like to watch two TV programs at once. While you watch the "main" channel you can watch the other channel in a small "window" in any corner of the screen. In that window might be a videotape, a computer game, satellite reception, videotext, even your home TV camera keeping an eye on the kids in the next room or the security system of the front door. This feature is being called PIP (picture-in-picture).

Throw in the Kitchen Sink

If all of this isn't enough for you, maybe what you want is an integrated system. Your television, turntable, videodisc, tuner, and receiver are all controlled by one wireless remote-control unit. You could control over a dozen different components from the palm of your hand. The commands you issue are echoed on a monitor that confirms your wishes. Already in the works is a system to allow you to control any of your electrical appliances through your television set. With small computer brains now being built into refrigerators and microwave ovens, not only will they talk to you ("Close the door, dummy"), but also you will be able to bark right back to them via your little key pad or home computer, through the TV monitor.

Now you can see why I said you need to do your homework before buying. Read a few magazine reviews, find out what is being offered, and come into the store knowing what features you want and how much you want to pay.

Whoever said "It's a jungle out there" must have just left a television showroom.

Spinning Wagon Wheels

Next time you're watching the stagecoach roll into town in old Westerns like *High Noon,* take a close look at the spinning wagon wheels. They appear to spin backward, don't they? In fact, in all the Westerns, when the bad guys or the Indians are chasing the stage, the wheel spokes always appear to spin backward, yet the coach keeps on rollin' along forward.

Ever wonder why this happens? It's an optical illusion produced by the movie camera and projector.

The first thing that has to be understood is that movie film is a series of still shots. Look at any piece of movie film and you can see the individual frames. The film projector in the theater runs these stills at twenty-four frames per second. In other words, when the projector is running, it is "blinking" on and off twenty-four times per second. It shoots the light through one frame, moves the film, then shoots the light again over and over, twenty-four times per second. So the screen is actually flickering, but your brain ignores the flickering, and you see a smooth-running movie.

However, the flickering does do tricks to spinning objects in the film. If the spokes were turning at the same rate that the film was flickering—twenty-four times per second—it would appear that the spokes were standing still, frozen in motion by the flashing light. Every time the wheel made one revolution, the projector would flash and show the spokes in the same place, so they would appear to be stationary.

But let's say the projector moves the film just a little faster than the wagon wheels are spinning. Then the spokes don't get quite back to their original positions, so they appear to have shifted backward. A whole series of these little shifts backward make the wheels appear to be spinning in reverse while the stagecoach is going forward.

Need to be convinced? Try this for yourself with your own television and a portable egg beater.

The TV serves as a flashing light source, like the projector. TV screens actually blink at thirty frames per second, a little faster than film frames. The egg beater will be like our wagon wheels, spinning quickly.

Turn on the egg beater and the TV. Hold the spinning beaters up in front of your eyes, look through the beaters at the TV screen, and what do you see? The beaters appear to be moving backward! If not, adjust the speed of the beaters. You can make them appear to stand still, move backward, or even move forward. The flashing TV light is creating the same optical illusion with the spinning egg beaters as it did with the wagon wheels.

Just one precaution: Make sure you've cleaned off the cake batter before attempting this sure-to-please demonstration.

The CD: It's the Pits

You've heard the powerful, lifelike notes produced by compact disc (CD) players, and you're ready to invest in one. Or you're waiting for the new Japanese digital tape machines to reach your local stereo store. But before banishing those *nouveaux* antique LP's to the attic, you'd like to know why compact discs sound so much better.

The answer is simple: digital recording. Encoding the music as a series of numbers is an ingenious solution to the many problems and shortcomings of conventional records and tapes.

The Shrinking Circle

Despite the advances in recording techniques, the old LP makes incredible demands on your turntable's needle (stylus) and the cutter that makes the grooves at the recording studio. As the needle moves from the outside of the record toward the inside, the speed of the record remains at a constant $33\frac{1}{3}$ revolutions per minute, but the distance the stylus travels shrinks. Grooves on the outside of the LP are much larger than grooves on the inside (actually, there is only *one* groove on each side of an LP, albeit a very long and winding one, but it is more convenient to think of an LP as having many grooves). Like drivers on a racetrack jockeying for the shorter inner lanes, the stylus travels a much shorter distance per revolution of the record—six inches per second—when it rides on the inner grooves than on the outer ones, sixteen inches per second. That means that the information—notes, music—contained on the inner grooves must be packed much closer together by the record cutter. Forcing more information into less room leads to a loss of space, which gobbles high-frequency notes. The cutting and playback styli simply can't follow the wavy lines fast enough to faithfully reproduce the music. (In the early days of LP's, recording engineers simply cut out the higher frequencies to make the music recordable, and in the process much of the music quality was lost.)

217

Low frequencies turn recording engineers' hair gray, too. Bass instruments recorded on both left and right stereo channels can literally cause the needle, in its valiant effort to faithfully track both sides of the groove at the same time, to jump out of the groove. As a result, bass notes were sacrificed for the good of the recording.

Advances in recording techniques—the use of computers—have helped alleviate the problems somewhat, but there is no easy way around the physical limitations of the LP, whose basic operating principle has changed little since Edison spoke the first recorded words of "Mary Had a Little Lamb" into the phonograph in 1877. Each LP copy pressed in plastic is not quite as good as the studio "master" recorded on tape or disc. Quality is lost in the duplicating process. The vinyl records are subject to warping and scratching. Heat makes them melt. Spilled food and drink render them all but unplayable. Recording an audiotape instead of an LP solves the problem of warping and scratching but creates problems of its own. Tape is inherently noisy. It introduces an element of "hiss" to all recordings, which can become especially annoying during quiet musical passages when the hiss may be louder than the music. Tape stretches with use, and the music can become noisier and distorted with repeated usage.

Is it possible to find a method of recording and playing music that eliminates the noise, distortion, and inaccuracies that plague the recording industry and as a bonus would increase the quality of the music being recorded? Where could such a solution be found? The answer is simple and elegant: computers. Computers must deal with billions and billions of pieces of information. They must be able to store and process immense amounts of detailed material and, most importantly, do it very *accurately*. A misplaced zero on an accounting ledger stored in a computer can mean the difference between boom and bust. So computers must be very fast, accurate, and reliable.

Why not take this ability and put it to work in the recording industry? That's exactly what's been done. Music played in the recording studio is immediately translated into the language of computers—"binary digits" or "bits"—and stored on tape or disc. These bits are then reproduced by the millions on small, round, silvery, five-inch discs called CD's.

The music on a CD is recorded as a digital code of numbers,

literally carved out by a laser beam. The laser burns away the surface, creating a series of "pits." The lengths of the pits and the spacing between them reflect the computer binary codes of ones and zeros. All this occurs on a fantastically small scale, too small for the naked eye to see. The laser beam in a CD must track a spiral three miles long, staying focused on a series of pits spinning at five hundred revolutions per minute, with a beam width just 1.7 microns (0.000068 inch) wide. Pick up a CD and see for yourself. No grooves can be seen—just a shiny, plastic-coated, metallic surface sandwiched between protective layers of clear plastic.

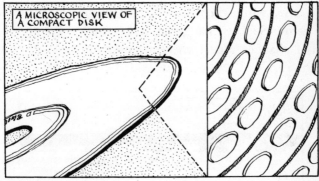

A MICROSCOPIC VIEW OF A COMPACT DISK

Illustration 82
A microscopic view of the pits on a compact disc. The pits are arranged in thousands of circular tracks spiraling inward, like grooves on a conventional LP. But these are not grooves. No needle ever touches them.

Numbers Don't Lie

In comparison with LP's, compact discs make noise and distortion so low as to be inaudible. The waveform of a complex audio sound, such as music, may look like the Rocky Mountains. The numerous hills and valleys coincide with the loudness of the music at different times. This mountainous terrain is impressed on your LP, and its close ride along these fluctuations is what makes life miserable for your stylus. The digital world does away with this problem by eliminating the rugged terrain and the needle. In digital recording the music is "sampled" many times—up to about forty thousand times per second—and the amplitude of the sound (the height of each peak) is assigned a number, with higher numbers for higher peaks. Each sample value—each number—is expressed as a binary code of ones and zeros and recorded in that form on magnetic tape or burned onto a disk by a laser beam.

In the recording process, the laser either burns away the surface, creating a "pit," or leaves the surface unscarred, producing a "land." On playback (see Illustration 83), a laser beam scans the spinning disc. Pits scatter the laser light while the lands strongly reflect it. The difference in reflected light intensity is detected by the CD player's computerized brain. The bright and dim reflections are converted to a binary code of ones and zeros. A pit-to-land or land-to-pit transition produces a one, while the length between the transitions creates a string of zeros. So the change in light intensity and the spacing between the changes reflect the computer binary code of ones and zeros.

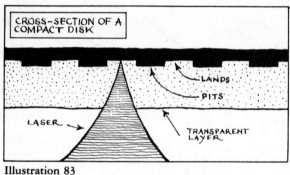

CROSS-SECTION OF A COMPACT DISK

LANDS

PITS

LASER

TRANSPARENT LAYER

Illustration 83

What are the advantages? There is very little wear and tear on the CD, so it may last forever. On the other hand, the very first time the needle touches a vinyl LP, the grooves begin to wear out. The worn-down peaks no longer accurately reflect the original recording, and the music is degraded ever so slightly with use and age. But consider digital recording. When the sound is recorded and read by light beam instead of the mechanical process of riding up and down peaks and valleys, the original information is protected. The distortion, noise, and tape dropouts unavoidable in analog records and tapes have no effect in the digital world. A number is a number; it can be only a one or a zero. As long as the laser can tell the difference between a pit and a flat spot, the number can be read, the information is not lost. And since nothing comes in contact with the "number" to wear it down—there is no needle or stylus in contact with the disc—the information is never degraded.

As for ruining the CD with scratches or food and drink, peanut butter and jelly can be easily washed or wiped off the CD's protective plastic coating. And since the encoded music is being continually processed by a tiny computer inside the CD player, any gaps in the recording created by scratches or tiny holes are filled in electronically by the CD's brain so that no annoying "pops" make their way to the speakers. Any imperfections in the recording or noise produced in the recording procedure are also ingeniously detected and corrected by the computer. Quiet passages are noise-free and hiss-free; there is almost dead silence. Music dramatically shifts from loud to soft and back again.

As a result, even the most inexpensive CD player will produce music comparable with the most expensive analog record and tape systems. Experts say CD is as close to hearing the master tape in your living room as you'll probably ever get.

Of course, the one drawback to CD's is that one cannot record directly onto them, as one can do with audio tape. However, new digital audio tape systems, or DAT's, are available that will do away with that limitation. As music is fed into the tape machines, it is digitized and stored on tape like conventional cassette recorders.

Not to be left behind, the video world has entered the laser disc market. The latest generation of compact discs contain snippets of video as well as audio. Called "CD-Vs" (the "V" stands for video), the discs comes in four sizes and can play from seventy minutes on a one-sided disc to sixty minutes a side on a two-sided disc.

White-Knuckling It or Window Seat, Please

Airline travel is hours of boredom interrupted by moments of stark terror.
—Al Boliska

Illustration 84

It's not unusual for the passengers on a jumbo jet to burst out in spontaneous applause when the wheels of the airliner touch down. Flying can be a frightening experience, and after a bouncing, stomach-turning ride through a thunderstorm at thirty thousand feet, few words can describe the joy of feeling terra firma beneath your feet. It doesn't matter how many times you've flown, how many hours you've logged in the air. There come times when you feel frightened, when you're positive the plane is not working just right, when you wonder what is making those strange noises you've seemingly never heard before. Thoughts turn to the inevitable "Will we make it?"

Everyone who flies has felt frightened at some point. Don't believe anyone who says he or she hasn't. I have, especially the time when the

222

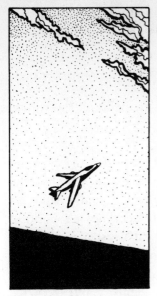

Illustration 85

plane was hit by lightning on the approach to La Guardia in New York. But through it all, I keep telling myself one thing: "I believe in the laws of physics." Like Peter Pan rousing the faithful to the aid of Tinker Bell, the laws of physics are my crutch in a crisis. They help get me through the worst stomach-turning situation, be it a harrowing thunderstorm or a red light on the pilot's instrument panel. I know that the plane stays in the air because of the laws of physics, and no matter how bad the weather is outside, those laws are not going to be violated. The plane should keep flying, barring a midair collision or gross human error.

So for all those who white-knuckle your way from one place to another, this chapter's for you. Presented for your approval, as Mr. Serling used to say, a short explanation of why an airplane flies, what goes on inside

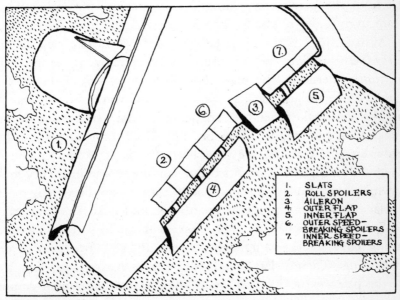

1.   SLATS
2.   ROLL SPOILERS
3.   AILERON
4.   OUTER FLAP
5.   INNER FLAP
6.   OUTER SPEED-
     BREAKING SPOILERS
7.   INNER SPEED-
     BREAKING SPOILERS

Illustration 86

the cockpit while you're nervously looking around, and a play-by-play explanation of what those funny little sounds and wing movements mean. There's lots of aircraft choreography going on outside your window, and by paying a little attention, you can enjoy the show instead of fearing it.

We will not try to explain Mr. Serling's gremlins on the wings.

Taking Off

Before the plane ever gets to the runway, the pilot and crew are working through a long checklist. They're checking whether the correct amount of fuel has been loaded (the pilot is actually handed a fuel slip by the person pumping the stuff), if the engines are working properly, whether the doors are shut tight. That's just a partial list, but the actual process is not unlike the countdown that precedes a space launch; and in fact, countdowns are really descendants of pilots' checklists. Even before the checklist is run through, you'll often see the pilot looking under the wings or poking around the fuselage. This is his or her own little way of making sure the ground crew hasn't missed anything.

Back to the plane. You'll notice that as the plane is being pushed back from the gate or is taxiing to the runway, unfamiliar noises scream from under the belly: The plane starts "whining."

What's that? Is the plane falling apart even before we take off? Quick! Get us back to the gate! Actually, an extra "wing" is sprouting out from underneath the plane (see Illustration 86). Don't get so excited. These are just the flaps, and the sound you heard was just the noise the flaps make when they are being moved. The flaps are crucial for taking off and landing. They provide extra lift for the plane at low speeds, so you'll always see (and hear) the flaps extending during takeoff and in the last moments before landing. Sometimes they make an awful racket, but it's a sound you'll learn to love because it means the plane is working properly and the ground is still not far away.

Some planes—like the 727—have a movable part on the front (leading edge) of the wing. The edge will "nose over" on takeoff and landing. These "slats" work together with the flaps; they also give the plane extra lift, and you'll know when the flight is coming to an end when you see the slats and flaps deployed. (If you should

224

be fortunate enough to pass through a cloud during takeoff or landing, you can watch the slats and flaps at work guiding the onrushing air. It's as if you, in window seat 22F, were sitting in a wind tunnel observing the moisture as it traces the contours of air flowing over the wing.)

So now you're taxiing into position. Did you catch the runway number painted on the pavement? It's a good landmark. The number indicates in what direction you'll be heading as you take off. Simply take the number and add a zero to it. Let's say our runway is 9. Adding a zero gives us 90. We are headed in compass direction of 90 degrees, which is exactly east. Let's say we're taking off on that runway but from the other end. What would the runway number be? (This is a trick question.) If you said 27, you'd be right because now you'd be heading west, which is 270 degrees. Subtracting the zero gives us runway 27. Notice that the two numbers—90 and 270—are 180 degrees apart—in opposite directions, as they should be. So runways change numbers depending on the direction in which you travel. Sometimes two runways will run alongside each other as they do at Los Angeles International Airport. Do they have the same numbers? You bet. Are the pilots confused? I hope not. One runway is designated "left," the other "right," so you might have runways 9L and 9R.

Okay, now we've waited out the logjam on the ground, the flaps and slats are down and the engines roaring. We're being pressed back into our seats as the plane accelerates and the pilots are working to get us off the ground. Takeoffs are probably the most frightening parts of flying. Who has never found himself talking to the plane, encouraging it to gain speed and take off safely? Takeoffs are riskiest because loss of power means the pilot has to brake quickly on the ground or continue to take off with reduced power. But nothing goes wrong on this flight. As we reach takeoff speed, the pilot pulls back on the yoke (the "stick" for you old-timers) and rotates the nose of the plane up into the air. The front of the plane leaves the ground first, the rest of it follows, and we're airborne.

What? Another frightening noise? It's only the sound of the landing gears (wheels) being retracted into the belly of the plane. All's well as we head off into the wild blue yonder.

Why the Wright Brothers Were Right

Yes, but what miracle of nature keeps us aloft? Remember our runway number? The direction in which we take off depends on the wind direction. We always want to fly *into* the wind. This is a strange idea. Why fight the wind when you want to gain speed? You wouldn't drive your car into the wind. Yes, but your car doesn't have wings (unless it was built in the early 1960s). The important part in getting the plane off the ground is not its speed relative to the ground but its speed relative to the air. Let's say a plane needs to get up to a hundred miles per hour (airspeed) to take off. These hundred miles per hour are measured relative to the wind. Let's say the plane is headed into a twenty-five-mile-per-hour wind. How fast would it need to travel down the runway to take off? Only seventy-five miles per hour. The twenty-five-mile-per-hour head wind plus the seventy-five-mile-per-hour ground speed add up to the hundred miles per hour needed to take off. So if you're a pilot, you want to head into the wind because it means easier takeoffs. Vice versa, if the plane has a twenty-five-mile-per-hour wind at its back, pushing, then it would need to take off at 125 miles per hour ground speed. Theoretically, if you were heading into a hurricane with winds of a hundred miles per hour, the plane would be standing still, making no headway relative to the ground, but "flying" all the same.

It's the speed of the plane relative to the air that makes it fly. But how does this happen? Obviously, the air must be pushing up on the wings more than gravity is pulling down on the plane. But how do you get the air pressure to push so much harder against the underside of the wing than the top? Look at a cross section of the wing, as in Illustration 87.

What do you notice about the wing? The topside is curved while the bottom is straight (just like the sail on a sailboat). Since the shortest distance between two points is a straight line, the distance from the front of the wing to the back is shortest on the bottom side of the wing. If you were air flowing over the wing, you would have to travel farther if you took the top route versus the bottom route. And in order for you (on top) to reach the back of the wing the same time

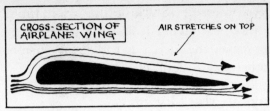

Illustration 87

an air particle on the bottom did, you would have to run much faster; you'd have more distance to cover in the same time. This means you would have to stretch yourself thin to make the same amount go farther. Air stretched thinly on top doesn't weigh as much or exert as much pressure as air packed thickly under the wing. So you'd have greater air pressure on the bottom of the wing, giving the plane "lift." You might look at it in a different way and say the thin air on top is "sucking up" the wing in the same way in which you suck up soda in a straw.

The plane also achieves lift because the leading edge of the plane is higher than the back (trailing) edge. No doubt when you were young and eager to experiment, you noticed this effect when riding in a fast-moving car. By holding your hand horizontally out the window with your fingertips pointed slightly up, your hand takes off like a plane. The hand imitates a stone skipping on a pond, and the wing owes some of its lift to the stone-skipping effect.

Remember the flaps and slats? During takeoff and landing, they give the plane added lift because they exaggerate the curved shape of the wing, and they also help the plane wing to skip on the air. So why not keep the flaps and slats deployed all the time? Too much wind resistance would result, called "drag." The flaps are not needed at high speeds; they would slow the plane down and could actually be ripped off the wings at the over-five-hundred-mile-per-hour speeds at which a jet airliner flies.

Sooner or later the plane will have to turn. Many people find the left or right roll unnerving but it's necessary, and for an interesting reason. The plane must bank or roll to make the turn since it has no roadway to grip. To begin the turn, little "flippers" near the middle of the wing—ailerons—flip up and down. To turn right, the right-wing aileron turns up. Stuck up like a sore thumb, the aileron meets the air head on, forcing it and the wing to "roll" and turn. Watch the pilot jockey the aileron up and down—first one way to roll, and then the other way to level off.

Highways in the Sky

Airliners don't fly just anyplace they want to. You have to stay on the road in your car, and a pilot has to stay on the road in the sky. There are highways in the sky with rules of the road as strict as those on the ground (without the highway patrolmen, of course). After takeoff, our airplane has been assigned an altitude, speed, and course to fly. The pilot has a set of maps in the cockpit that show him (or her) the "road." Scattered across the country are over nine hundred radio transmitters, called radio ranges, that send out radio beams. These are the "signposts." The pilot flies from one transmitter to another, lining his plane up with a particular beam or course he or she wants to fly. The radio signals tell the pilot the distance to the next post and how far away the plane is.

On the ground, air traffic controllers are following the plane, which shows up as a "blip" on a radar screen. If you're lucky enough to be on a flight on which you can listen in on the air traffic channel, you'll be able to hear air traffic controllers talking with all the planes—directing them to change altitude, speed, and direction; jockeying them into different positions on the air highways.

The pilots on the plane can do something you can't: They can take their hands off the wheel while driving. The automatic pilots are so intelligent that all the human needs do is enter the course, speed, and altitude and the plane literally flies itself. Some of the newest planes, like the Boeing 757, can actually land themselves. (I tried my hand at "landing" a 757 in a flight simulator, which never really leaves the ground. Needless to say, I had the dubious honor of crashing the jumbo jet into the runway at Boston's Logan International Airport. The flight simulator is so realistic in imitating rough flying conditions and crash landings that pilots come away with sweaty palms and black-and-blue marks.)

Touchdown

The landing, for me, is the most interesting part of the flight. For the pilots, the boring job of watching the autopilot fly the plane is over, and there is lots to do. If the pilot is landing by instruments, instruments on board are tuned into two radio beacons: One tells the plane if it's flying left or right of the runway, the other if it's above or below the runway. The copilot calls out air speed and altitude to the pilot. The pilot has taken over control of the plane and is fighting wind gusts and turbulence to keep the plane headed for the tiny runway in the distance.

The high-pitched whine of the flaps and slats extending usually signals the start of the approach. This is the best time to observe the action. With the flaps fully extended, it almost appears as if they're going to screw themselves off the wing (you can see the screw mechanisms turning).

If the plane is coming in too high or too fast, the pilot may elect to use the wing "spoilers" or speed brakes. These are those flat wing plates that flip up and look like matchbook covers. The spoilers do just what their name implies: spoil the shape of the wing so that smooth airflow is interrupted and the wing loses lift and speed. The plane drops quickly—"dumping altitude"—and the rapid descent can be felt in the pit of your stomach. (I can usually tell when an old aircraft carrier pilot is at the controls because some like to make believe they are back in the Navy and nose-dive into a landing. They drop altitude quickly and make good use of the speed brakes.) On the ground, the spoilers act as an air brake to slow the plane on the runway quickly following touchdown.

With flaps down, spoilers up, and slats deployed, the wing appears to be naked: You can see right through the bare hydraulic plumbing to the ground. To the novice flier the sight is a bit unnerving, but to the savvy wing jockey, this is what flying is all about. Watching the pilot fight the crosswinds or dump altitude quickly by deft shifts of the ailerons and spoilers lends a bit of excitement to the humdrum of a smooth-running flight. (Computers are increasingly taking the place of pilots in moving the wing parts. They react quicker and make the flight smoother.)

Night flying is the most beautiful. There are very few sights to match that of a major metropolitan city like New York, Chicago, Boston, or San Francisco when viewed from a few thousand feet above sea level at night. The blinking of yellow streetlights between the trees, the long expanse of lit highway, the tall buildings rising majestically are all given a holiday

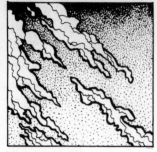

Illustration 88

glow, as if Christmas has come early. Then off in the distance comes a light display like no other: a rainbow of blue and yellow, red, and white. It's the airport. Airport lights are color-coded. Runways are outlined in white or yellow. Taxiways are bordered in blue. Red lights mark dangerous obstructions that must be avoided. Sometimes the lights are mounted many inches off the ground. Why is that done? If you live in the North, you know: snow.

One of the most important sets of lights tells the pilot whether he is coming in too high or too low. These are the approach lights at the ends of the runway. The aircraft has to be landed at an approach angle of three degrees. If the plane is above this angle, the lights will show up white; below this angle they appear red. By watching these colors shift, the pilot can judge the angle of approach.

The pilot lines up with the center of the runway, aims for the numbers and just a few feet above ground, lowers the tail, and raises the nose—"flares out." The rear wheels touch down first, then the front wheels, speed-braking spoilers flip up, and the jet engines give out a tremendous roar as the thrust of the engines is reversed to brake the plane. (On propeller-driven planes, the pitch of the propellers is reversed, not the direction of spin of the blades themselves.) The plane taxis slowly to the gate, and you have another successful flight under your belt. Years of aeronautical design and training have paid off in a safe flight. Now, if only the same kind of dedication and effort could go into delivering your luggage.

TRIVIA: The little black box is actually orange. The flight recorder is housed in an orange box, a color more easily detected amid the rubble of an airplane crash. Inside, airspeed, altitude, and vertical acceleration are scratched on a piece of stainless steel tape about as thick as aluminum foil. The voice recorder, contained in a separate orange "black box," is a thirty-minute tape recording of the last conversation in the cockpit plus any transmissions that have been made on the radio.

Wings bend a lot. Don't be afraid when you see the wings on the plane "flapping" in the wind. They are made to bend. In tests, the wings of a Boeing 747 have been bent twenty-five feet without breaking. It is normal for them to bend eight to nine feet in the course of a routine flight. If the wings don't bend, then you're in trouble. Bending relieves stress; wings that don't bend, break. Wings are made to support two and a half times the weight of the aircraft.

How important are the slats on takeoff? The crash of an American Airlines DC-10 at Chicago's O'Hare International Airport in 1979 occurred when the left engine and pylon ripped off the wing and fell to the ground. Losing an engine is not necessarily a fatal event. But in tumbling off the wing, the engine cut a hydraulic line that controlled the slats, leaving one useless. Without the lift, the left wing could not "fly." Unbalanced lift on the right wing, still flying, caused the plane to roll over and crash.

Index

Index

Index

Index

Index

Index